*Elastic Constants
and Their Measurement*

Elastic Constants
and Their Measurement

EDWARD SCHREIBER, Ph.D.
Professor of Earth and Environmental Sciences
Queens College, City University of New York
Visiting Senior Research Associate
Lamont-Doherty Geological Observatory of
Columbia University

ORSON L. ANDERSON, Ph.D.
Professor of Geophysics
Institute of Geophysics and Planetary Physics
University of California at Los Angeles

NAOHIRO SOGA, Ph.D.
Associate Professor of Industrial Chemistry
Kyoto University

McGRAW-HILL BOOK COMPANY
New York   St. Louis   San Francisco   Düsseldorf   Johannesburg
Kuala Lumpur   London   Mexico   Montreal   New Delhi   Panama
Paris   São Paulo   Singapore   Sydney   Tokyo   Toronto

Library of Congress Cataloging in Publication Data

Schreiber, Edward, date.
 Elastic constants and their measurement.

 Includes bibliographical references.
 1. Elasticity. I. Anderson, Orson, L., joint author.
II. Soga, Naohiro, joint author. III. Title.
TA418.S37   620.1'1232   73-19986
ISBN 0-07-055603-2

Copyright © 1973 by McGraw-Hill, Inc. All rights reserved.
Printed in the United States of America. No part of this
publication may be reproduced, stored in a retrieval system,
or transmitted, in any form or by any means, electronic,
mechanical, photocopying, recording, or otherwise, without
the prior written permission of the publisher.

1234567890 KPKP 7654

*The editors for this book were Jeremy Robinson and Stanley E. Redka,
the designer was Naomi Auerbach, and its production was supervised
by George E. Oechsner.
Printed and bound by The Kingsport Press.*

Contents

Preface   ix

1. *The Elastic Moduli* 1

   1.1 Introduction 1
   1.2 The Elastic Moduli 2
        1.2.1 Young's Modulus E and Poisson's Ratio σ 2
        1.2.2 The Bulk Modulus B and the Shear Modulus G. 3
   1.3 Constants Arising from Acoustics 5
   1.4 Nonisotropic Elastic Constants 7

2. *Elasticity in Crystals* 9

   2.1 Introduction 9
   2.2 Relation of the Velocity of Sound to the Elastic Properties 9
        2.2.1 Stress and Strain 9
        2.2.2 Elastic Constants 13
   2.3 Crystal Symmetry and Elastic Properties 14
        2.3.1 Operator Notation 14
        2.3.2 Matrix Reduction for the Crystal Classes 17
   2.4 Plane Wave Propagation 25

2.5	Calculation of the Isotropic Bulk and Shear Moduli	29
2.6	Adiabatic Character of Acoustic Measurements	31
	REFERENCES	34

3. The Determination of Velocity of Propagation 35

3.1	Introduction	35
3.2	Transit-Time Measurements	36
3.3	Pulse-Echo Methods	46
3.4	Acoustic Interferometry	59
	3.4.1 Phase-Comparison Methods	59
	3.4.2 Pulse Superposition	68
	3.4.3 Transducers and Coupling Materials	73
	3.4.4 Measurements at High Pressure and High Temperature	75
	REFERENCES	79

4. Dynamic Resonance Method for Measuring the Elastic Moduli of Solids 82

4.1	Introduction	82
4.2	Vibration of Cylindrical and Rectangular Bars	83
4.3	Torsional Vibration	84
4.4	Flexural Vibration	88
	4.4.1 Cylindrical Rods	90
	4.4.2 Rectangular Bars	90
4.5	Longitudinal Vibration	91
4.6	Measuring System	98
4.7	Identification of the Vibration Mode	100
4.8	Calculation of Elastic Moduli from Resonant Frequencies	101
4.9	The Effect of Orientation upon the Relation of Young's Modulus and Shear Modulus to the Elastic Compliances	102
4.10	Measurements at High Temperatures	107
4.11	Vibration of a Wire or Bar Clamped at one End	111
4.12	Other Methods of Employing Resonance	114
	4.12.1 Piezoelectric effect	115
	4.12.2 Electromagnetic effect	117
	4.12.3 Electrostatic effect	119
4.13	Problems Related to Internal Friction	120
	REFERENCES	122

5. Resonant-Sphere Methods for Measuring the Velocity of Sound 126

5.1	Introduction	126
5.2	Free Oscillations of a Sphere	127
5.3	Specimens and Their Preparation	132
5.4	Equipment	133
5.5	Method of Operation	135
5.6	Analysis of Data	136
	REFERENCES	141

6. Indirect Methods of Estimating Elastic Constants 143

- 6.1 Introduction 143
- 6.2 Estimating Elastic Constants 144
  - 6.2.1 Estimating the Shear Velocity from the Specific Heat 144
  - 6.2.2 Using the Debye Temperature and Bulk Modulus to Find the Isotropic Elastic Moduli 147
- 6.3 Estimation of the Bulk Modulus from the Volume 149
  - 6.3.1 Pure Compounds 149
  - 6.3.2 Solid Solutions 156
- 6.4 Use of the Infrared Reflection of a Diatomic Solid to Determine Bulk Modulus 157
- 6.5 Estimating the Temperature Derivatives of Elastic Moduli 159
- 6.6 Estimating the Pressure Derivative of the Bulk Modulus 163
  - 6.6.1 Estimation from the Grüneisen Parameter 163
  - 6.6.2 Estimation of dB/dP from the Repulsion Exponent in the Born Potential 163
  - 6.6.3 Determination of Bulk Modulus B and dB/dP from High-Pressure P-V Data 166
- REFERENCES 166

7. The Pressure and Temperature Derivatives of Elastic Constants and Thermodynamic Functions 168

- 7.1 Introduction 168
- 7.2 Pressure Dependence of Elastic Constants of Cubic Crystals 170
- 7.3 Pressure Dependence of Elastic Constants of Hexagonal Crystals . . . 177
- 7.4 Assorted Formulas for Young's Modulus, Shear Modulus, and Poisson's Ratio 183
- 7.5 Adiabatic Isothermal Transformations 187
- REFEENCES 189

Index 191

Preface

There has been an expanded interest in the measurement of the elastic properties of solids by using ultrasonic techniques, for they have the very desirable aspect of being a form of nondestructive test. A parallel and salutory development of electronic equipment has resulted in a variety of techniques for the performance of these measurements. They range in precision from a percent to a hundredth of a percent at ambient conditions and under conditions of high temperature, high pressure, or both simultaneously. Part of the impetus behind these advances came from the needs of materials scientists, where, as new alloys, composites, plastics, and ceramic material underwent development and design, a knowledge of their elastic properties was essential. Since these new materials were often designed for exotic applications, it became necessary to determine these properties, with precision, under unusual environmental conditions. This, together with the effect of the high cost in developing the materials, imposed the added requirement that methods be found that were capable of using smaller and smaller specimens. Today, it is routine in some laboratories to perform measurements on samples with dimensions of only a couple of millimeters, and measurements have been performed upon specimens of lunar material only fractions of a millimeter in size. There exists an extensive literature concerning the physics of elasticity and those static methods that have been used in its measurements. The dynamic methods, because of their nondestructive character and experimental simplicity, have also gained wide acceptance. In spite of this, there are few texts that are devoted to the dynamic methods of determining elastic constants. This book is designed to help fill that gap.

In writing this book it was our purpose to bring together a description of the various methods that have been developed to measure elastic properties. We have grouped them according to the nature of the measurement. Thus, Chapter 3 concerns itself with methods involving traveling waves, while Chapters 4 and 5 consider methods employing standing waves. In Chapter 6, methods of estimating these properties are discussed for those instances in which it is not feasible to perform the measurements. A presentation of the measurement of the elastic constants would be incomplete without a discussion of what is being measured and the extent of the complexities involved, a province of Chapter 2, while Chapter 7 is concerned with the problems of data reduction for the more difficult experiments performed under elevated temperatures and pressure, and the fundamental relationships between elastic properties and thermodynamics. This book is aimed at workers in physics, the geological sciences, the material sciences (ceramics, metallurgy, polymers, etc.) and the engineering disciplines where for a variety of reasons the determination of the elastic properties of materials is important; it is also directed at the upper class undergraduate level or graduate level of training, depending somewhat on the discipline involved.

In considering the variety of dynamic methods of measurement available to the experimentalist, some comments concerning the choice of particular techniques may be of worth, since such a decision involves a commitment of time and money. In general the main criterion is that of obtaining the needed precision at the least cost and expenditure of effort. If the demand on precision is not high, the straightforward pulse-transmission methods may be used. System components are simple and usually readily available in most laboratories. Specimens are easily prepared and measurements at ambient conditions can be routinely made. Measurements with the sample under pressure are also easily made with this method. If it is desirable to determine the temperature variation of elastic constants, the resonance techniques described in Chapters 4 and 5 can be employed to advantage. Components are only somewhat more complex than in the case of pulse transmission, but the requirements on specimen geometry are more stringent. Sphere resonance is amenable to use with very small samples, however, and may offer singular advantages if sample size is a critical consideration. Where measurements of the highest precision are needed, the more sophisticated pulse transmission or acoustic interferometric methods discussed in Chapter 3 may be most useful. These are also of particular advantage where the variation of elastic properties is to be determined under conditions of elevated pressure and temperature simultaneously. As the changes in elastic properties due to these parameters are small, the very great precision obtainable is advantageous and often outweights the compensating disadvantages of severe requirements on specimen preparation and higher equipment costs. The ultimate choice depends on the nature of the measurement to be made and the investment in effort, time and money that can be made. No sweeping recommendations can be made; rather it is necessary to carefully evaluate ones individual requirements and select that technique which most satisfies the needs.

This volume was developed while the authors were all associated with the Lamont-Doherty Geological Observatory of Columbia University, Palisades, N.Y., with the sponsorship of the United States Air Force under Contract AF 33(615)-1700. The contract was initiated under the direction of Air Force Materials Laboratory, Air Force Systems Command; Captain Peter J. Marchiando served as project monitor.

The project was initiated to investigate and develop a variety of methods for determining the mechanical properties of small specimens. This volume stems from that effort. It is intended to serve those involved in the measurement of the elastic properties of solids.

Few books are ever the sole creations of their authors, and this one is no exception. We owe a considerable debt of gratitude to a number of people who read the volume at its various stages of development and provided very useful and critical comments. Among these are Drs. Relva Buchanan, Robert C. Liebermann, Donald E. Scheule, and John B. Wachtman, Jr. The authors are also indebted to Mrs. Lorraine Wiley and Mrs. Marjorie Wassermann, who struggled through the handwritten manuscripts and earlier drafts to provide the typed manuscript; to Miss Kazuka Nagao, who provided the excellent draftsmanship; to Miss Paloma Hormess, who did yeoman service in proofreading through various stages; and B. Charlotte Schreiber, who in addition to assisting in the later, critical stages of proofreading, undertook the arduous task of indexing.

Edward Schreiber
Orson L. Anderson
Naohiro Soga

CHAPTER ONE

The Elastic Moduli

1.1 Introduction

The elastic properties of solid materials are of considerable significance to both science and technology. Their measurement yields information concerning the forces that are operative between the atoms or ions comprising a solid, information that is fundamentally important in interpreting and understanding the nature of bonding in the solid state. Because the elastic properties describe the mechanical behavior of materials, their measurement is also important for purposes of engineering design. The choice of the most appropriate material for a particular application requires a knowledge of its mechanical properties. Hence these parameters are needed in determining design criteria in order to avoid failure of an object in use.

Elastic properties are those properties which govern the behavior of a material subjected to stress over a region of strain where the material behaves elastically. When a material is subjected to a stress it will deform; that is to say, it will become strained. The way in which a material deforms over a large range of applied stress is shown in Fig. 1.1.

Over the region of stress level up to σ_y, the strain responds linearly. If the stress is reduced, the strain diminishes reversibly, and upon removal of the stress, the strain goes to zero. At a sufficiently high level of stress, the strain is no longer simply linear, and removal or reduction of the stress does not result in reversible strain; a permanent deformation results. The material is behaving plastically. Further stressing of the material causes additional plastic deformation and finally rupture. This behavior is typical of most solids except those which are exceptionally brittle and which fracture in the region of elastic behavior.

The proportionality constant relating stress and strain in the region of elastic behavior is the elastic constant, or more precisely, elastic modulus. In technological research on materials the elastic moduli are often treated as empirical numbers used to describe rather complex properties (in terms of first principles), and a certain methodology has arisen which, while adequate for certain applications, departs from the traditional point of view concerning elastic moduli.

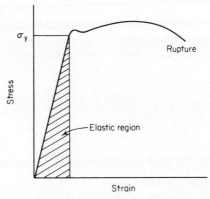

Fig. 1.1 Typical stress-strain curve. The slope of the linear portion of the curve is the elastic modulus; the area under the curve, the work done in straining the material elastically.

1.2 The Elastic Moduli

1.2.1 Young's Modulus E and Poisson's Ratio σ.

Among the most elementary notions in considering the strength of a material is Young's modulus. In engineering applications, materials are employed in such a manner as to make this property an important design parameter, for this modulus relates a *unidirectional* stress to the resultant strain. Young's modulus is taken as the measure of the resistance to traction along the axis of a thin bar or rod and is sometimes called simply the *elastic modulus*, a name that should be avoided, since it implies that there is only one elastic modulus. Such a modulus is illustrated in Fig. 1.2a. A rod or bar is shown subjected to a uniaxial stress (tensile in the example in the figure) and is strained as a consequence. The resulting strain is manifested by an elongation in the direction of the applied stress and a decrease in the diameter of the rod. The change in length divided by the initial length $\Delta l/l$ is the strain, and the ratio of the stress to the strain is Young's modulus, another way of saying that it is the slope of the elastic portion of the stress-strain curve shown in Fig. 1.1.

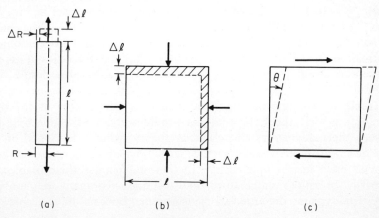

Fig. 1.2 Illustrating the sort of deformation under various types of applied stress. (a) Uniaxial load; (b) hydrostatic compression; (c) pure shear.

In addition to elongating in the direction of the axially applied stress, the rod simultaneously narrows. This change in dimension in a direction that is perpendicular to the direction of the applied stress develops from an interaction of the strain components that are generated within the material as a consequence of being stressed, principally as a volume-conserving strain. An additional elastic modulus is needed to describe this behavior. This is represented by Poisson's ratio, which is defined as the ratio of change in dimension in lateral direction to the change in length $2\Delta R/\Delta l$.

Many technological problems are often adequately handled with a knowledge of Young's modulus and Poisson's ratio alone. These are only useful, however, if the material is both homogeneous and isotropic. They have therefore been used with great success in the engineering disciplines, since most engineering materials are homogeneous and isotropic. As interest in materials becomes more directed toward the application of single crystals, account must be taken of their nonisotropic character.

1.2.2 The Bulk Modulus B and the Shear Modulus G. From a fundamental viewpoint, Young's modulus and Poisson's ratio are not the simplest ones even for isotropic materials. While technologically elastic moduli arise as empirical constants in Hooke's law as applied to large bodies, a greater significance is implied in their appearance as the second derivative of the elastic energy density with respect to strains. For isotropic materials two fundamental properties arise as a result of subjecting the material to two different states of stress. These are (1) the bulk modulus, where the strains perpendicular to the stress directions are all equal (the case of response to hydrostatic pressure), and (2) the shear modulus, where the strains perpendicular to the directions of stress are everywhere zero (pure shear). These are illustrated (for the two-dimensional case) in Fig. 1.2b and c. The strain in the case of the bulk modulus is the ratio of the change in volume to volume $\Delta V/V$; the measure of the shear strain is the amount of deformation through the angle θ. The moduli are the ratios of applied stress to the strain in the regions where the material behaves elastically.

There are a number of reasons why the bulk and shear moduli are preferable to Young's modulus and Poisson's ratio. One of the most important is that the bulk modulus provides the best connecting link between thermodynamics and elasticity theory, which, developed on the basis of atomistic viewpoints such as lattice dynamics, is concerned with behavior on a microscopic scale. In thermodynamics the variables are ordinarily pressure, temperature, and volume. In many relationships the thermodynamic definition of the isothermal bulk modulus occurs:

$$B = -V\left(\frac{dP}{dV}\right)_T \tag{1.1}$$

which connects the variables of pressure P and volume V. In pure shear, on the other hand, the deformation involves no change in volume, only a change in shape.

There are other important reasons for considering the bulk and shear moduli as having a fundamental significance since the behavior of the bulk modulus is quite different from that of the shear modulus. In "stiff" materials, the values of both bulk and shear moduli are high, whereas in "supple" materials the value of the bulk modulus remains high and the value of the shear modulus is markedly decreased. Temperature has a marked effect on the shear modulus, but a much lesser effect on the bulk modulus, and at the melting point the value of the former goes to zero, while the value of the latter may be only slightly affected. Further, the application of very high stresses affects the value of the shear modulus. That is to say, above a certain level of shear stress, the body is no longer elastic but begins to flow. On the other hand, under a pure compressional stress, a body remains elastic up to enormous pressures.

A further reason why B and G are important is that they occur as a natural consequence whenever one develops theories of elastic constants employing an atomistic approach, a procedure which involves a proposed theory of the forces of attraction between atoms, and a method to estimate the variation of these forces with strain. The attractive forces postulated are often spherically symmetrical, and the internal energy of the system is resolved into a part which represents dilatation (compression) and other parts which represent distortion (shear). In these approaches, the bulk modulus B always appears, as do one or more shear moduli. On the other hand, the meaning of Young's modulus is not very clear in theories using an atomistic approach.

1.3   Constants Arising from Acoustics

Although B and G are fundamental elastic constants which arise from very basic considerations, they do not always coincide with those elastic constants which are simplest to measure. When elastic properties are determined from the measurement of accoustic wave propagation through a material, then other moduli are involved. To obtain the latter a condition of dynamic equilibrium is defined where the forces are functions of the stress, and the associated strain components are functions of displacements within the solid. The forces are set equal to the inertial forces, which show how the displacements vary with time according to Newton's laws, and are described by a set of wave equations. For isotropic bodies, geometric considerations appropriate to the direction of wave propagation result in a reduction to two simple equations. The first equation is given by the condition that the displacements perpendicular to the wavefront direction are zero, and this yields the longitudinal wave equation. The second equation is given by the condition that the displacement along the direction of the wavefront is zero, which yields the shear wave equation for the isotropic case. In one dimension these reduce to

$$\frac{d^2u}{dt^2} = \frac{L}{\rho}\frac{d^2u}{dx^2} \quad \text{longitudinal}$$

$$\frac{d^2u}{dt^2} = \frac{G}{\rho}\frac{d^2u}{dx^2} \quad \text{shear} \tag{1.2}$$

where $\rho =$ density
$u =$ displacement

These two kinds of waves are easy to generate in the laboratory and to distinguish from each other so that, from the *experimental* view, the constants L and G are of considerable importance. The constant of the shear wave is precisely the same G (shear modulus) discussed before. The longitudinal modulus L, determined from the velocity of propagating a longitudinal wave, is neither the bulk modulus B nor Young's modulus E,

TABLE 1.1 The Connection between Elastic Constants of Isotropic Bodies*

B	E	λ	σ	$L = \rho v_l^2$	$G = \rho v_s^2$	How obtained
$\lambda + \dfrac{2G}{3}$	$G\dfrac{3\lambda + 2G}{\lambda + G}$	...	$\dfrac{\lambda}{2(\lambda + G)}$	$\lambda + 2G$	...	λ is always calculated
...	$9B\dfrac{B-\lambda}{3B-\lambda}$	...	$\dfrac{\lambda}{3B-\lambda}$	$3B - 2\lambda$	$\dfrac{3(B-\lambda)}{2}$	
...	$\dfrac{9BG}{3B+G}$	$B - \dfrac{2G}{3}$	$\dfrac{3B-2G}{2(3B+G)}$	$B + \dfrac{4G}{3}$	...	Static measurement of B and G
$\dfrac{EG}{3(3G-E)}$	...	$G\dfrac{E-2G}{3G-E}$	$\dfrac{E}{2G} - 1$	$G\dfrac{4G-E}{3G-E}$	...	Static measurement of E and G; dynamic (resonance) measurement of E and G
...	...	$3B\dfrac{3B-E}{9B-E}$	$\dfrac{3B-E}{6B}$	$3B\dfrac{3B+E}{9B-E}$	$\dfrac{3BE}{9B-E}$	Static measurement of B; dynamic (resonance) measurement of E
$\lambda\dfrac{(1+\sigma)}{3\sigma}$	$\lambda\dfrac{(1+\sigma)(1-2\sigma)}{\sigma}$	...	...	$\lambda\dfrac{1-\sigma}{\sigma}$	$\lambda\dfrac{1-2\sigma}{2\sigma}$	
$G\dfrac{2(1+\sigma)}{3(1-2\sigma)}$	$2G(1+\sigma)$	$G\dfrac{2\sigma}{1-2\sigma}$	...	$G\dfrac{2-2\sigma}{1-2\sigma}$	...	Static or dynamic measurement of G and σ (e.g., using sphere resonance)
...	$3B(1-2\sigma)$	$3B\dfrac{\sigma}{1+\sigma}$	...	$3B\dfrac{1-\sigma}{1+\sigma}$	$3B\dfrac{1-2\sigma}{2+2\sigma}$	Static measurement of B and σ
$\dfrac{E}{3(1-2\sigma)}$	...	$\dfrac{E\sigma}{(1+\sigma)(1-2\sigma)}$	...	$\dfrac{E(1-\sigma)}{(1+\sigma)(1-2\sigma)}$	$\dfrac{E}{2+2\sigma}$	Static or dynamic measurement of E and σ
$\rho(v_l^2 - \tfrac{4}{3}v_s^2)$	$\dfrac{3v_s^2 v_l^2 - 4v_s^4}{v_l^2 - v_s^2}$	$\rho(v_l^2 - 2v_s^2)$	$\dfrac{v_l^2 - 2v_s^2}{2(v_l^2 - v_s^2)}$	...	...	Dynamic (velocity) measurement of v_l and v_s

* Adapted from an article by Prof. Francis Birch, and appeared originally in the *Journal of Geophysical Research*, vol. 66, No. 7, p. 2206, 1961.

although it is often confused with these two constants. The former was seen to relate a volumetric strain to a hydrostatic stress, and Young's modulus, a strain to a flexural stress, whereas the longitudinal modulus relates the strain to a longitudinally applied stress. The longitudinal modulus is expressed in a variety of forms. For reasons involved with simplicity in expressing the mathematical forms of elasticity, it is most often expressed as

$$L = \lambda + 2G = \rho v_l^2 \tag{1.3}$$

where $\lambda =$ Lamé constant
$G =$ shear modulus
$v_l =$ longitudinal velocity

In all, six elastic moduli, B, G, E, σ, L, and λ, are defined.

The elastic moduli of an isotropic body may be expressed in a variety of ways. It can be shown that any two are sufficiently complete to describe elastic behavior, and that any elastic modulus may be expressed in terms of any other pair. Table 1.1 lists these interconnecting relationships between the isotropic elastic moduli.

1.4 Nonisotropic Elastic Constants

A solid which is isotropic as far as its elastic constants are concerned is one in which the elastic moduli are not functions of direction. This may be checked experimentally by determining if the sound velocities depend upon the direction of propagation, or if the measurement of Young's modulus, using a bar or rod cut in one direction from a sample, is found to be different from Young's modulus as determined with a rod or bar cut along a different direction in the same sample. This is not to be equated with optical isotropy, for isotropy in elasticity is not the same as isotropy in optical mineralogy. A cubic crystal, such as NaCl, is not isotropic with regard to elastic constants although it is isotropic with regard to its optical properties. In fact, the measured sound velocities vary greatly with direction in many cubic crystals.

The kinds and number of elastic constants that one must consider in nonisotropic bodies are described in the following chapters. It is shown that a cubic crystal has three independent elastic constants which are given as c_{11}, c_{44}, and c_{12}. The nature of the subscripts and their meaning are given in Chap. 2. In comparing elastic constants of a cubic crystal to an isotropic material, the important point to remember is that c_{11} represents a longitudinal modulus and not Young's modulus. The cubic moduli c_{44} and c_{12} correspond to the shear modulus $G = \rho v_s^2$ in particular ways. In fact, there are two shear moduli which arise in the wave equations for isotropic materials; however, they are equal in magnitude, and it is customary to state that there is but one shear modulus in isotropic materials. One of the shear moduli in a cubic crystal is designated c_{44}, and the other is $\frac{1}{2}(c_{11} - c_{12})$. When the value of c_{44} approaches closely the value $\frac{1}{2}(c_{11} - c_{12})$ for all directions in the material, a cubic crystal is said to be nearly isotropic. The condition of an isotropic cubic crystal is approached in the cases of garnet and tungsten.

The constants c_{44} and c_{12} have an important relationship when one considers them from an atomistic point of view. If one postulates central forces between the atoms in a solid, as is done, for example, in the ionic model of solids, and further specifies that every atom is a crystallographic center of symmetry, it can be shown that the shear constants of the lattice are prescribed by the *Cauchy* relation,

$$c_{12} = c_{44}$$

This condition is sometimes used as a method of differentiating covalent from ionic bonding. Both NaCl and MgO are cubic crystals; the Cauchy relation is closely observed in NaCl, but it is deviated from considerably in the case of MgO, which has a strong covalent character. In cubic solids, the Cauchy relation is satisfied only when Poisson's ratio is exactly 0.25.

The existence of two shear constants in cubic crystals results from the two different ways which a solid may be strained and still have its volume conserved. The first is to change the angle between the sides of a cell, but keep the length of the sides constant. The second is to keep the angles between the sides constant, but to arrange that the amount of volume lost by squeezing parallel faces is gained by expanding other sets of parallel faces. In the first way, modulus c_{44} is involved. In the second way, the shear strain involves two orthogonal extensions and this creates two constants, one of which is c_{11} (that is, pushing along 1 and extending along 1), and the other c_{12} (that is, pushing along 1 and extending along 2). While neither c_{11} nor c_{12} by itself is a shear, their combination amounts to a shear.

In an isotropic solid, both methods of straining the solid are equivalent, for only a rotation is needed to lead from one definition to the other. This situation is prescribed by the condition that the shear modulus G is invariant with direction.

In considering crystallographic structures with symmetry lower than that of the cubic system, the number of elastic constants increases. These complications are considered in more detail in Chap. 2. The main features which we have discussed remain the same. There are simply more independent longitudinal moduli to be measured and more independent shear moduli to be measured; and in the case of the shear moduli, more Cauchy relationships are involved.

CHAPTER TWO

Elasticity in Crystals

2.1 Introduction

The measurement of the velocity with which sound waves are propagated through a solid has become an important means of determining the elastic constants of materials. This is true because techniques employing high-frequency acoustic waves are both experimentally more convenient and inherently more precise than are the older static methods. Since the problem of determining the velocity cannot be divorced from the nature of the crystals themselves, the relationships between their elastic properties and the sound velocity are reviewed.

2.2 Relation of the Velocity of Sound to the Elastic Properties

2.2.1 Stress and Strain. The elastic constants are defined in terms of the response of a crystal to an applied stress. For the general case of an anisotropic crystal, we must consider all the components of strain that arise as a result of the application of a general stress. These topics are reviewed in summary by Huntington[1] and in detail by Nye.[2] Stresses and strains are tensor properties of the second rank and therefore require nine numbers to completely specify them. For the case of stress, consider generalized forces on the face of a unit cube within a stressed body. These may be resolved into stresses parallel to and normal to an orthogonal axial reference frame, as shown in Fig. 2.1, for normal compression.

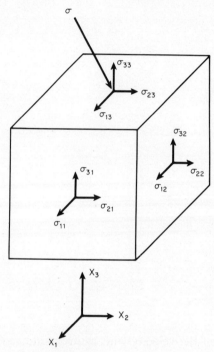

Fig. 2.1 Stress components arising from generalized stress σ (shown on one face only); the right-hand axial system is shown below.

There are, in general, nine components of stress denoted as σ_{ij}. The convention for notation is that i refers to the axis to which the stress component is parallel, and j refers to the cube face at which the stress is applied. Thus σ_{11} refers to a stress parallel to the X_1 axis ($i=1$) and applied on the face normal to the X_1 direction ($j=1$), σ_{12} is the stress component parallel to the X_1 axis applied on the X_2 cube face, and so on. The stress components for which $i=j$ are the *normal* components of stress, and those for which $i \neq j$ are the *shear* components of stress. The sign convention is chosen so that positive values of σ_{11}, σ_{22}, σ_{33} each represent a tensile stress, negative values a compressive stress. When the directions of the applied stresses are also chosen parallel to an appropriate reference frame, the shear stresses become zero, i.e., when σ_1, σ_2, and σ_3 are parallel to σ_{11}, σ_{22}, and σ_{33}, respectively.

The deformation which a body undergoes when subjected to a stress system is the strain, and it is not the simple displacement of points in a body which is of concern, but rather their motions *relative to one another*. This is an important point, because the former can describe the motion of the whole unstrained body relative to some axial frame of reference. Thus, when *all* the points in the body move a distance δ in the $+X_1$ direction, this motion describes *translation* and *not* strain. On the other hand, if a point p in a body is translated an amount δ in the $+X_1$ direction, and a nearby point within the body p' moves δ' in the $+X_1$ direction (where $\delta \neq \delta'$), then the body is strained, even though as a whole it may not have been moved. The work done (i.e., the strain energy) is given by $\sigma dA \cdot \delta$. The concern here is with the motion of points within the body rather than motions of the body as a whole.

Implicit in the discussion above is that the amount of strain which a given point undergoes within a body is a function of its position coordinate within the body (note that $\delta \neq \delta'$). This is the basis of the definition for strain, i.e.,

$$\varepsilon_{ij} = \frac{\partial U_i}{\partial X_j} \qquad i,j = 1, 2, 3 \tag{2.1}$$

where $U_i =$ displacement of a point within the body
$X_j =$ a coordinate of the reference frame

For the three-dimensional case, there are nine strains which can be specified for the various combinations of i, j. Of these nine, three are compressional or tensional strains $(i=j)$ (elongations or foreshortening), and six are shear strains $(i \neq j)$. The measure of the shear strains in terms of the deformation of a unit element in the body is shown in Fig. 2.2. If we refer only to axes X_1 and X_2, then the shear strain is defined for strains in the X_1, X_2 plane. As shown in Fig. 2.2, the distortion consists

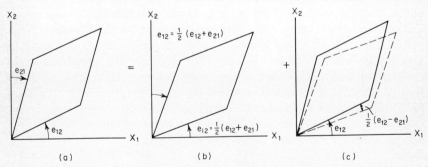

Fig. 2.2 (*a*) Generalized strain is composed of two parts; (*b*) the symmetric tensor corresponds to the deformation $\tfrac{1}{2}(e_{12} + e_{21})$, which is the true strain; (*c*) the antisymmetric tensor corresponds to a rotation through the angle $\tfrac{1}{2}(e_{21} - e_{21})$. (Reproduced through permission of Prof. T. F. Nye, from "Physical Properties of Crystals," published by Clarendon Press, Oxford, 1957.)

of a strain and a rotation. Only those components which represent the strain $(e_{12}+e_{21})$ and not those that represent rotation $(e_{21}-e_{12})$ are of interest here. We will use the notation $\varepsilon_{ij} = \frac{1}{2}(e_{ij}+e_{ji})$ to represent the components of strain.

In summary, the stress and strain tensors may be expressed as arrays of nine numbers which represent the three components of stress and strain along the directions of the coordinate axes. The stress tensor represents a force which can be applied along any arbitrary direction of the crystal and is called a field tensor, as is the strain tensor. The tensor which measures the crystal property is a *matter* or *property* tensor. These are the elastic constants.

The stress and strain tensors are of second rank and may be represented by the following arrays:

The *stress* tensor is

$$\begin{bmatrix} \sigma_{11} & \sigma_{12} & \sigma_{13} \\ \sigma_{21} & \sigma_{22} & \sigma_{23} \\ \sigma_{31} & \sigma_{32} & \sigma_{33} \end{bmatrix} \quad (2.2a)$$

In the absence of body torques $\sigma_{ij} = \sigma_{ji}$, so the above array is simplified to six independent stress components. If, in addition, the principal strains coincide with the reference axes, the array simplifies further, because all the σ_{ij} will then be zero, for $i \neq j$.

The *strain* tensor

$$\begin{bmatrix} e_{11} & \frac{1}{2}(e_{12}+e_{21}) & \frac{1}{2}(e_{13}+e_{31}) \\ \frac{1}{2}(e_{12}+e_{21}) & e_{22} & \frac{1}{2}(e_{23}+e_{32}) \\ \frac{1}{2}(e_{13}+e_{31}) & \frac{1}{2}(e_{23}+e_{32}) & e_{33} \end{bmatrix} \quad (2.2b)$$

is also symmetric ($\varepsilon_{ij} = \varepsilon_{ji}$) and represents the displacement of a point in the deformed body. If the principal strains $i = j$ conform to the Cartesian reference frame, then the shear strains vanish ($\varepsilon_{ij} = 0$). It is important to realize that the property of the σ_{ij} or ε_{ij} going to zero with the appropriate choice of reference directions exists because these are field tensors. The matter tensors (i.e., elastic constants) will not vanish simply as a result of an arbitrary rotation of axis because they represent the properties of the crystal, which are independent of the location or orientation of the reference frame. Finally, it should be pointed out that the elements of the array for stress or strain represent the components of either stress or strain in three mutually perpendicular directions, so that with stress, for example, we can write the equations in component form as

$$\begin{aligned} \sigma_1 &= \sigma_{11}X_1 + \sigma_{12}X_2 + \sigma_{13}X_3 \\ \sigma_2 &= \sigma_{21}X_1 + \sigma_{22}X_2 + \sigma_{23}X_3 \\ \sigma_3 &= \sigma_{31}X_1 + \sigma_{32}X_2 + \sigma_{33}X_3 \end{aligned} \quad (2.3)$$

where $\sigma_1, \sigma_2, \sigma_3$ are the three generalized stresses.

2.2.2 Elastic Constants.

In order to relate a stress to the strain it causes in a material body, it is necessary to know a priori how they are connected or else to assume from experience a simple, and therefore restrictive, relationship. Theory being inadequate to provide the former, the latter approach is chosen, and it is assumed that the strain is simply proportional to the stress. The assumption, which, of course, is Hooke's law, is restricted to small strains and the requirement that upon stress release the body resumes its unstressed shape, that is, behaves elastically. We have, then, that stress is proportional to strain

$$\sigma_{ij} \propto \varepsilon_{kl} \tag{2.4}$$

The constant of proportionality is a material property and, in general, depends on its direction within the material. Further, since it is a number that links two second-rank tensors, it is, itself, a tensor of the fourth rank. This tensor is the elastic stiffness c_{ijkl} (note subscripting). Its inverse is the elastic compliance s_{ijkl} and connects the strain to the stress. The relation between these constants and stress and strain is

$$\sigma_{ij} = \Sigma c_{ijkl} \varepsilon_{kl} \tag{2.5a}$$

$$\varepsilon_{ij} = \Sigma s_{ijkl} \sigma_{kl} \tag{2.5b}$$

It is these stiffnesses or elastic compliances which are the properties that are determined in measurements upon the elastic behavior of solids.

To simplify writing the stiffness or compliance tensor, an abbreviated notation is employed which expresses the fourth-rank tensor with two subscripts rather than four.[2] This notation simplifies the means of expressing the compliances or moduli in matrix form. The operation for conversion from the tensor to the matrix notation is

Tensor	11	22	33	23,32	31,13	12,21
Matrix	1	2	3	4	5	6

The reduction in the array from 9 to 6 results from the symmetry of the stress and strain tensors ($ij = ji$). Expressed in compressed notation, the stress and strain tensors are

$$\sigma_i = \sum_{j=1}^{6} c_{ij} \varepsilon_j \tag{2.6a}$$

and

$$\varepsilon_i = \sum_{j=1}^{6} s_{ij} \sigma_j \tag{2.6b}$$

and i goes from 1 to 6, so that written in expanded (component) form there are six equations. These are

$$\begin{aligned}
\sigma_1 &= c_{11}\varepsilon_1 + c_{12}\varepsilon_2 + c_{13}\varepsilon_3 + c_{14}\varepsilon_4 + c_{15}\varepsilon_5 + c_{16}\varepsilon_6 \\
\sigma_2 &= c_{21}\varepsilon_1 + c_{22}\varepsilon_2 + c_{23}\varepsilon_3 + c_{24}\varepsilon_4 + c_{25}\varepsilon_5 + c_{26}\varepsilon_6 \\
\sigma_3 &= c_{31}\varepsilon_1 + c_{32}\varepsilon_2 + c_{33}\varepsilon_3 + c_{34}\varepsilon_4 + c_{35}\varepsilon_5 + c_{36}\varepsilon_6 \\
\sigma_4 &= c_{41}\varepsilon_1 + c_{42}\varepsilon_2 + c_{43}\varepsilon_3 + c_{44}\varepsilon_4 + c_{45}\varepsilon_5 + c_{46}\varepsilon_6 \\
\sigma_5 &= c_{51}\varepsilon_1 + c_{52}\varepsilon_2 + c_{53}\varepsilon_3 + c_{54}\varepsilon_4 + c_{55}\varepsilon_5 + c_{56}\varepsilon_6 \\
\sigma_6 &= c_{61}\varepsilon_1 + c_{62}\varepsilon_2 + c_{63}\varepsilon_3 + c_{64}\varepsilon_4 + c_{65}\varepsilon_5 + c_{66}\varepsilon_6
\end{aligned} \quad (2.7)$$

and a similar set of six equations is written for the compliances s_{ij}. The c_{ij} or s_{ij} comprise the 6×6 matrix of the elastic constants. As noted earlier, c_{ij} and s_{ij} are the inverse of each other, but it should be pointed out that since each forms a 6×6 array, the inverse operation is not a simple one. To go from the c_{ij} to the s_{ij} matrix, or the reverse, requires *inversion of the matrix*. This numerical calculation is not a simple operation for a 6×6 matrix and is more conveniently handled with a computer. The matrices and their inverses are summarized in Mason (chap. 3).[4]

2.3 Crystal Symmetry and Elastic Properties

2.3.1 Operator Notation. The symmetry properties of crystals have the effect of reducing the number of independent elastic constants. The higher the crystal symmetry, the fewer the independent terms that exist in the matrix. For example, the triclinic system, with symmetry operations 1 or $\bar{1}$, has no symmetry insofar as the elastic properties are concerned. The matrix consists of 21 independent elastic constants and not 36. This reduction is due to the fact that terms ij and ji are equal because the matrix is symmetric, as seen from the physical arguments concerning the stress and strain tensors discussed above. The stress tensor is symmetric because the body is in dynamic equilibrium. The strain tensor is the symmetric part of the displacement tensor; the antisymmetric part represents a pure rotation and *not* a strain.[2] The introduction of a single twofold symmetry axis (monoclinic system, class 2) reduces the number of independent elastic constants to 13, or almost by half. Imposing only the minimum symmetry of the cubic system (class 23) reduces the number of independent elastic constants to 3, and this number is reduced to 2 for an isotropic medium.

In order to understand the reason for the reduction in the matrix of the elastic constants, it is useful to start first with a brief review of the point symmetries and their operations. When appropriate directions are chosen as axes, all crystals may be grouped on the basis of their macroscopic morphology into one of 32 crystal classes (symmetry point groups) which are subgroups among 7 crystal systems. Although it is common practice to combine the systems of threefold and sixfold axial symmetry and refer to 6 crystal systems, the distinction is made here because, insofar as many property tensors are concerned, the two are not equivalent. In Table 2.1 the symmetry properties of crystals are summarized. In going from the cubic to the triclinic system, the symmetry of the axes which define the system is systematically reduced. The first reduction is one of scale, all axes no longer having the same definition for the unit of length. This

TABLE 2.1 Symmetry Operators for the 32 Crystal Classes

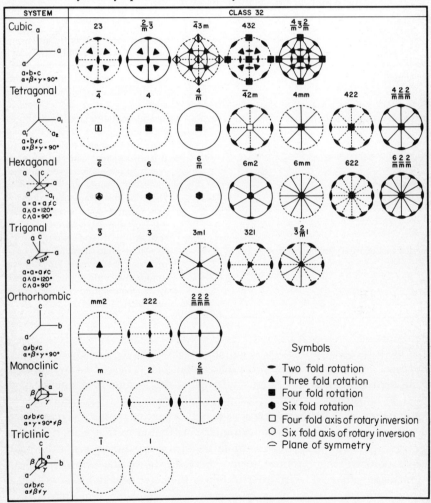

reaches its limit in the orthorhombic system, and to decrease the symmetry further, one must forego orthogonality; the triclinic system exhibits the least possible symmetry, with all axial lengths and interaxial angles arbitrary.

The crystal classes comprise the combinations of the point symmetry operations which result in distinct groupings consistent with the definition of the crystal system to which they belong. The point symmetry operations are those which operate on a point in space in such a manner as to produce a symmetrically equivalent point, and by repetition of the operation, the original point is once again attained. These operations fall into several groups which are described below.

Operation/Symbol	Description
Inversion, $\bar{1}$	A point is inverted through the origin—all its coordinates change sign.
n-fold axis of rotation, $n = 1, 2, 3, 4, 6$	Equivalent points are located by a rotation of $2\pi/n$. A fivefold rotation does not exist in crystals because it is inconsistent with the requirements of translational symmetry.
Mirror planes, m	Equivalent points are located by reflection in a plane normal to an axis (and as with a mirror, the left- or the right-handedness is reversed by the reflection).
n-fold axis of rotary inversion $\bar{n}$, $\bar{n} = \bar{1}, \bar{2}, \bar{3}, \bar{4}, \bar{6}$	An n-fold rotation followed by an inversion stepwise. Only $\bar{4}$ is unique and could not be expressed as a combination of the other symmetry elements.

The list, while not exhausting all possibilities, is sufficient for our purpose. As seen in Table 2.1, each crystal class is described by one or a combination of these operations. In using the symmetry elements of a class to construct a special crystallographic form, there is frequently a redundancy, with further operations merely repeating points. To produce the general form $\{hkl\}$, all the symmetry attributed to that class need not be used. However, application of all the symmetry operations of the next lower class will not produce forms characteristic of the symmetry class immediately above it. In order to follow the notation which describes the the symmetry of each class compactly, the following conventions are adhered to.

Cubic System. In the cubic system, the first operator position refers to the three principal (a_1) axes; the second operator position to the four body diagonals (a_2); the third operator position to the six face diagonals (a_3). Mirror planes are perpendicular to the axis to which the symmetry operator position refers, and combinations are shown as a fraction, $4/m$ (to be read as a fourfold axis of rotation with a reflection plane normal to the fourfold axis).

Tetragonal System. The first operator refers to the c axis, the axis of highest symmetry. The second operator refers to the two a_1 axes, and the third to the a_2 axes.

Trigonal and Hexagonal. The first position refers to the c axis, the second operator to the a axis, and the third to the a_1 axes.

Orthorhombic. The first, second, and third operator positions refer to the three crystallographic axes a, b, and c, in that order. The c axis is chosen as the longest of the three, the a axis the shortest.

Monoclinic. Two orientations are in use, so it is necessary to note both of them. In the first, the crystal is so oriented that the operator refers to the b crystallographic axis. In the alternative orientation of the crystal, the operator refers to the c axis.

Triclinic. The maximum symmetry exhibited by a triclinic crystal is a center of symmetry. (A onefold axis of rotation is zero symmetry, because the operation produces no new points which are symmetrically equivalent.)

A symmetry class denoted by $4/m\ \bar{3}\ 2/m$ (cubic hexoctahedral) means that *all* three a_1 axes are fourfold and have mirror planes normal to them, *all* four body diagonals are threefold axes of rotary inversion, and *all* six face diagonals are twofold axes with mirror planes normal to them. This follows from the meaning of symmetry—that equivalent directions or axes are indistinguishable from one another. This immediately suggests the basis for the reduction of the terms in the property matrix by imposing the point symmetry of the crystal.

2.3.2 Matrix Reduction for the Crystal Classes. To find the reduction of the matrix in a systematic fashion, it is useful to resort to the following approach. We can consider an orthogonal reference frame X_1, X_2, X_3, defined by a right-handed coordinate system, and align these with the crystallographic axes to the extent that the axial symmetry allows.

Using the symmetry operations of each class, we can choose to apply these either to the crystallographic axes a, b, c or to the reference axes X_1, X_2, X_3. This is possible because the property itself is invariant and independent of any initial orientation of the reference frame. Hence, rotating the reference frame relative to the crystallographic axes by applying the appropriate symmetry operators is equivalent to rotating the crystallographic axes relative to the reference frame.

We choose the latter because, by preserving the property of orthogonality, the problem of dealing with the axial transformations which result from the application of the symmetry operation is simplified. The only situations which have to be considered separately are the classes of the trigonal and hexagonal systems. This is so because, in these systems, the symmetry operations involve transformations where the axes are rotated other than in integral multiples of 90°.

The transformation of one set of axes into another is found by determining the direction cosines of the new axis (X') in terms of the old (X). Thus we may write in component form (a_{ij} are the direction cosines)

$$X_1' = a_{11}X_1 + a_{12}X_2 + a_{13}X_3$$
$$X_2' = a_{21}X_1 + a_{22}X_2 + a_{23}X_3 \qquad (2.8)$$
$$X_3' = a_{31}X_1 + a_{32}X_2 + a_{33}X_3$$

or more compactly

$$X_i' = a_{ij}X_j \qquad \text{summation is on } j$$

This simplifies considerably if the old axes are transformed to the new axes by rotations of 90°, since the direction cosine is then unity. For a counterclockwise rotation of 90° about X_3, we have $X_1 \to X_2'$ and $X_2 \to -X_1'$. Thus the properties associated with the X_1 axis are associated with the X_2' axis after rotation. The components of X_1 on X_1' and X_1 on X_3' are zero, hence $a_{11} = a_{13} = 0$. The matrix of the transformation is simply

$$a_{ij} = \begin{pmatrix} 0 & 1 & 0 \\ \bar{1} & 0 & 0 \\ 0 & 0 & 1 \end{pmatrix} \qquad (2.9)$$

For a rotation of 180°, the direction cosines of the axes normal to the rotation axis are -1, which is the same as inverting a point on the normal axis through the center, and from this we note that the operation of $\bar{1}$ inverts all the axes. A mirror plane inverts only the axis normal to it. These transformations provide the basis for determining the reduction of the elastic constant matrices by imposing the condition of point group symmetries for all crystal classes except the trigonal and hexagonal. We can now develop the reduced matrices of the elastic constants.

The simplest way of accomplishing the transformations of axes is to use the tensor notation. The fourth-rank tensor is defined by the transformation

$$T_{ijkl} = a_{im}a_{jn}a_{ko}a_{lp}T_{mnop} \qquad (2.10)$$

where T_{ijkl} is the tensor referred to the new axis and T_{mnop} the tensor referred to the original axis; the a terms are the sets of transformation coefficients (direction cosines) for each subscript pair. Since a fourth-rank tensor requires four subscripts, the transformation requires four sets of direction cosines. We illustrate Eq. (2.10) by applying it first to the triclinic system.

Triclinic System; Class 1. In the triclinic system, pedial class, the axes transform as $X_1 \to X_1'$, $X_2 \to X_2'$, $X_3 \to X_3'$ (really, nothing happens) and all the direction cosines are unity. Therefore, $T_{ijkl} = T_{mnop}$ and the 6×6 matrix of the 21 independent elastic constants remains unchanged.

Triclinic System; Class $\bar{1}$. In the basal class of the triclinic system, the axes transform as $X_1 \to -X_1'$, $X_2 \to -X_2'$, $X_3 \to -X_3'$, and all the transformation coefficients (a_{ij}') are -1 (see above). From Eq. (2.10) we have

$$T_{ijkl} = (-1)^4 T_{mnop} = T_{mnop}$$

The center of symmetry operator does not change the matrix of elastic constants either, and for both classes of the triclinic system there is no reduction in the number of independent elastic constants. From the foregoing, we conclude the elastic constants are centrosymmetric, as are *all* fourth-rank tensors. The presence of a center of symmetry will therefore not reduce the number of independent constants. Any reductions that arise must therefore develop from the other symmetry elements that may be present.

Monoclinic System; Class 2. In the prismatic class we have (in the usual orientation) the $b = X_2$ axis as the symmetry axis. Application of this symmetry element results in $X_1 \to -X_1'$, $X_2 \to X_2'$, $X_3 \to -X_3'$, and the coefficients of the transformation are

$$a_{ij} = \begin{pmatrix} \bar{1} & 0 & 0 \\ 0 & 1 & 0 \\ 0 & 0 & \bar{1} \end{pmatrix}$$

We now take the elastic constants tensor (say the stiffness) and transform each $c_{mnop} \to c_{ijkl}'$ term-by-term according to Eq. (2.10). This leads to nine sets of nine equations; using the above matrix to obtain the values of a_{ij}, we have

$$\begin{aligned}
c_{1111}' &= a_{11} a_{11} a_{11} a_{11} c_{1111} = (-1)^4 c_{1111} = c_{1111} \\
c_{1112}' &= a_{11} a_{11} a_{11} a_{21} c_{1111} = 0 \\
c_{1113}' &= a_{11} a_{11} a_{11} a_{31} c_{1111} = 0 \\
c_{1121}' &= a_{11} a_{11} a_{21} a_{11} c_{1111} = 0 \\
c_{1122}' &= a_{11} a_{11} a_{21} a_{21} c_{1111} = 0 \\
c_{1123}' &= a_{11} a_{11} a_{21} a_{31} c_{1111} = 0 \\
c_{1131}' &= a_{11} a_{11} a_{31} a_{11} c_{1111} = 0 \\
c_{1132}' &= a_{11} a_{11} a_{31} a_{21} c_{1111} = 0 \\
c_{1133}' &= a_{11} a_{11} a_{31} a_{31} c_{1111} = 0
\end{aligned} \quad (2.11)$$

The total number of terms that have to be dealt with in this way are $(9 \times 9)\,(21) = 1{,}701$ terms. The term-by-term procedure is clearly cumbersome, but it suggests a simpler one.[5] We note that the c transforms to the c' according to the product of the direction cosines a_{ij}. We therefore give *these* our attention. Referring to the subscripts, we note that in the suffix notation, the a_{ij} transforms the axis as

$$1 \to -1,\; 2 \to 2,\; 3 \to -3$$

Then in four-suffix notation, the *product* of *pairs* of a_{ij} results in

$$11 \to 11,\; 22 \to 22,\; 33 \to 33,\; 23 \to -23,\; 13 \to 13,\; 12 \to -12$$

which in the two-suffix matrix notation becomes (see Sec. 2.2.2)

$$1 \to 1,\; 2 \to 2,\; 3 \to 3,\; 4 \to -4,\; 5 \to 5,\; 6 \to -6$$

Thus, by applying the transform to the suffix notation itself, a very simple scheme is generated for the matrix transformation. The c_{ij} transforms to the c'_{ij} as follows, writing the new term in the *position* of the old:

$$\begin{matrix}
c'_{11} & c'_{12} & c'_{13} & -c'_{14} & c'_{15} & -c'_{16} \\
 & c'_{22} & c'_{23} & -c'_{24} & c'_{25} & -c'_{26} \\
 & & c'_{33} & -c'_{34} & c'_{35} & -c'_{36} \\
 & & & c'_{44} & -c'_{45} & c'_{46} \\
 & & & & c'_{55} & -c'_{56} \\
 & & & & & c'_{66}
\end{matrix}$$

We see that all the constants transform into themselves except for the eight which *only* change sign. These must be zero since the only number for which $a_{ij} = -a_{ij}$ is zero. The resulting matrix is given by

$$\begin{matrix}
c_{11} & c_{12} & c_{13} & 0 & c_{15} & 0 \\
 & c_{22} & c_{23} & 0 & c_{25} & 0 \\
 & & c_{33} & 0 & c_{35} & 0 \\
 & & & c_{44} & 0 & c_{46} \\
 & & & & c_{55} & 0 \\
 & & & & & c_{66}
\end{matrix}$$

Monoclinic; *Classes* m, $2/m$. The domatic class, with only a reflection plane, inverts only the b axis, so

$$X_1 \to X'_1,\; X_2 \to -X'_2,\; X_3 \to X'_3$$

which in the four-suffix notation is

$$11 \to 11,\; 22 \to 22,\; 33 \to 33,\; 23 \to -23,\; 13 \to 13,\; 12 \to -12$$

and we note that the effect of a mirror plane normal to an axis is the same as having the axis exhibit a twofold symmetry. Therefore, all three classes of the monoclinic system have the same matrix.

Orthorhombic; *Classes mmm, 222, 2/m 2/m 2/m*. From the equivalence of the mirror plane and twofold axis symmetries determined for the monoclinic system, we see that all three orthorhombic classes will be expressed by the same matrix. Further, we can start with the matrix for the monoclinic system and need apply only the additional symmetry to it. Applying the twofold symmetry operation first about the X_1 axis, we have

$$X_1 \rightarrow X_1', X_2 \rightarrow -X_2', X_3 \rightarrow -X_3'$$

or $\quad\quad 11 \rightarrow 11, 22 \rightarrow 22, 33 \rightarrow 33, 23 \rightarrow 23, 13 \rightarrow -13, 12 \rightarrow -12$

and $\quad\quad 1 \rightarrow 1, 2 \rightarrow 2, 3 \rightarrow 3, 4 \rightarrow 4, 5 \rightarrow -5, 6 \rightarrow -6$

Applying these to the pertinent elastic constants, we see that $c_{15} = c_{16} = c_{25} = c_{26} = c_{35} = c_{36} = c_{45} = c_{46} = 0$. Of these, all but c_{15}, c_{25}, and c_{35} were zero from before. If we now apply the third twofold symmetry element, about X_3, we have

$$X_1 \rightarrow -X_1', X_2 \rightarrow -X_2', X_3 \rightarrow X_3'$$

or $\quad\quad 11 \rightarrow 11, 22 \rightarrow 22, 33 \rightarrow 33, 23 \rightarrow -23, 13 \rightarrow -13, 12 \rightarrow 12$

and $\quad\quad 1 \rightarrow 1, 2 \rightarrow 2, 3 \rightarrow 3, 4 \rightarrow -4, 5 \rightarrow -5, 6 \rightarrow 6$

This leads to $c_{14} = c_{15} = c_{24} = c_{25} = c_{34} = c_{35} = c_{46} = c_{56} = 0$, and inspection shows that no new terms are zero. The matrix for all classes in the orthorhombic system therefore is

$$\begin{matrix} c_{11} & c_{12} & c_{13} & 0 & 0 & 0 \\ & c_{22} & c_{23} & 0 & 0 & 0 \\ & & c_{33} & 0 & 0 & 0 \\ & & & c_{44} & 0 & 0 \\ & & & & c_{55} & 0 \\ & & & & & c_{66} \end{matrix}$$

Tetragonal; *Classes 4, $\bar{4}$, 4/m*. From the considerations of the triclinic system, we found that the matrix of the elastic constants is centrosymmetric. It follows that classes 4 and $\bar{4}$ should be expressed by the same matrix. We anticipate the result of adding a mirror plane normal to the fourfold axis and include 4/m. The symmetry operator 4 applied to X_3 yields

$$X_1 \rightarrow X_2', X_2 \rightarrow -X_1', X_3 \rightarrow X_3'$$

or $\quad\quad 11 \rightarrow 22, 22 \rightarrow 11, 33 \rightarrow 33, 23 \rightarrow -13, 31 \rightarrow 32, 12 \rightarrow -21$

and $\quad\quad 1 \rightarrow 2, 2 \rightarrow 1, 3 \rightarrow 3, 4 \rightarrow -5, 5 \rightarrow 4, 6 \rightarrow -6$

Applying this to the matrix terms, we find that $c_{11}=c_{22}$, $c_{13}=c_{23}$, $c_{44}=c_{55}$, $c_{16}=-c_{26}$, and that $c_{14}=c_{15}=c_{24}=c_{25}=c_{34}=c_{35}=c_{36}=c_{45}=c_{46}=c_{56}=0$. The matrix, therefore, has the form

$$\begin{matrix} c_{11} & c_{12} & c_{13} & 0 & 0 & c_{16} \\ & c_{11} & c_{13} & 0 & 0 & -c_{16} \\ & & c_{33} & 0 & 0 & 0 \\ & & & c_{44} & 0 & 0 \\ & & & & c_{44} & 0 \\ & & & & & c_{66} \end{matrix}$$

If we now apply the mirror operator normal to X_3, we have $X_1 \rightarrow X_1'$, $X_2 \rightarrow X_2'$, $X_3 \rightarrow -X_3'$, which gives the same result as in the orthorhombic system. There we found that this operator leads only to having $c_{14}=c_{15}=c_{24}=c_{25}=c_{34}=c_{35}=c_{46}=c_{56}=0$. However, this result is already included in the reduction due to the fourfold symmetry. The class $4/m$, therefore, is expressed by the same matrix as classes 4 and $\bar{4}$.

Tetragonal; Classes 4mm, $\bar{4}2m$, 422, $4/m\ 2/m\ 2/m$. From the preceding discussions regarding the effect of a center of symmetry, where we found the equivalence of a twofold axis and of a mirror plane, and of a twofold or fourfold axis combined with a mirror plane, we conclude that these four crystal classes are all expressed by a single matrix of the elastic constants. If we start with the matrix arrived at from operation of the fourfold axis, and if we now apply, say, the twofold symmetry to the X_1 and X_2 axes, we have, respectively, for axis X_1

$$X_1 \rightarrow X_1',\ X_2 \rightarrow -X_2',\ X_3 \rightarrow -X_3'$$

and for axis X_2

$$X_1 \rightarrow -X_1',\ X_2 \rightarrow X_2',\ X_3 \rightarrow -X_3'$$

or for axis X_1

$$11 \rightarrow 11,\ 22 \rightarrow 22,\ 33 \rightarrow 33,\ 23 \rightarrow 23,\ 31 \rightarrow -31,\ 12 \rightarrow -12$$

and for axis X_2

$$11 \rightarrow 11,\ 22 \rightarrow 22,\ 33 \rightarrow 33,\ 23 \rightarrow -23,\ 31 \rightarrow 31,\ 12 \rightarrow -12$$

Then for axis X_1

$$1 \rightarrow 1,\ 2 \rightarrow 2,\ 3 \rightarrow 3,\ 4 \rightarrow 4,\ 5 \rightarrow -5,\ 6 \rightarrow -6$$

and for axis X_2

$$1 \rightarrow 1,\ 2 \rightarrow 2,\ 3 \rightarrow 3,\ 4 \rightarrow -4,\ 5 \rightarrow 5,\ 6 \rightarrow -6$$

which leads to $c_{14}=c_{15}=c_{24}=c_{25}=c_{34}=c_{35}=c_{36}=c_{45}=c_{46}=c_{56}=0$. These terms were already zero due to the effect of the fourfold axis of symmetry. In addition, the preceding requires that $c_{16}=c_{26}=-c_{26}$.

This can hold only if $c_{16} = c_{26} = 0$. Thus the matrix for these crystal classes is

$$\begin{matrix} c_{11} & c_{12} & c_{13} & 0 & 0 & 0 \\ & c_{11} & c_{13} & 0 & 0 & 0 \\ & & c_{33} & 0 & 0 & 0 \\ & & & c_{44} & 0 & 0 \\ & & & & c_{44} & 0 \\ & & & & & c_{66} \end{matrix}$$

Cubic System; All Classes ($23, 2/m3, 43m, 432, 4/m\overline{3}2/m$). We first note that we can divide these classes into two groups: those with six twofold axes and those with three fourfold axes. The first group, therefore, starts with the matrix for the orthorhombic system; the second, with the matrix of the tetragonal system. Considering the first two classes, we add the symmetry operator of the threefold rotation axis (body diagonal) which transforms as

$$X_1 \rightarrow X_3', X_2 \rightarrow X_1', X_3 \rightarrow X_2'$$

or $\quad 11 \rightarrow 33, 22 \rightarrow 11, 33 \rightarrow 22, 23 \rightarrow 12, 31 \rightarrow 23, 21 \rightarrow 13$

and $\quad 1 \rightarrow 3, 2 \rightarrow 1, 3 \rightarrow 2, 4 \rightarrow 6, 5 \rightarrow 4, 6 \rightarrow 5 \rightarrow 4$

By inspection, we see that this reduces the orthorhombic matrix by having $c_{11} = c_{22} = c_{33}$; $c_{12} = c_{13} = c_{23}$; and $c_{44} = c_{55} = c_{66}$. The same operator reduces the tetragonal matrix so that it is identical with the above reduction. Additional application of symmetry operators yields no further reduction. The final matrix is

$$\begin{matrix} c_{11} & c_{12} & c_{12} & 0 & 0 & 0 \\ & c_{11} & c_{12} & 0 & 0 & 0 \\ & & c_{11} & 0 & 0 & 0 \\ & & & c_{44} & 0 & 0 \\ & & & & c_{44} & 0 \\ & & & & & c_{44} \end{matrix}$$

and only three constants are required to describe the elastic properties of cubic materials.

Isotropic Materials; Isotropy results in a further reduction of the independent elastic constants to two. This may be shown by a transformation about the original reference axis of 45°. The matrix for isotropic materials is

$$\begin{matrix} c_{11} & c_{12} & c_{12} & 0 & 0 & 0 \\ & c_{11} & c_{12} & 0 & 0 & 0 \\ & & c_{11} & 0 & 0 & 0 \\ & & & \tfrac{1}{2}(c_{11} - c_{12}) & 0 & 0 \\ & & & & \tfrac{1}{2}(c_{11} - c_{12}) & 0 \\ & & & & & \tfrac{1}{2}(c_{11} - c_{12}) \end{matrix}$$

These summarize the elastic matrices for all the crystal classes which have only orthogonal symmetry elements. The scheme used to develop these matrices operates on the suffix notation because the transformation between the new and old axis took a simple form. For the trigonal and hexagonal case, these transformations are not so simple because they involve rotations of either 120° or 60° about the X_3 axis. The transformation matrices are shown below.

1. Trigonal   2. Hexagonal

$$a_{ij} = \begin{pmatrix} -\tfrac{1}{2} & \tfrac{1}{2}\sqrt{3} & 0 \\ -\tfrac{1}{2}\sqrt{3} & -\tfrac{1}{2} & 0 \\ 0 & 0 & 1 \end{pmatrix} \qquad a_{ij} = \begin{pmatrix} \tfrac{1}{2} & \tfrac{1}{2}\sqrt{3} & 0 \\ -\tfrac{1}{2}\sqrt{3} & \tfrac{1}{2} & 0 \\ 0 & 0 & 1 \end{pmatrix}$$

This complication necessitates the evaluation of the transformed elastic constant matrix by actual matrix multiplication. This is a procedure too lengthy to repeat here, and so the resulting matrices are listed. The reader may wish to carry out the necessary operations to obtain the result given below.

Trigonal; Classes $\bar{3}$, 3

$$\begin{matrix} c_{11} & c_{12} & c_{13} & c_{14} & -c_{15} & 0 \\ & c_{11} & c_{13} & -c_{14} & c_{15} & 0 \\ & & c_{33} & 0 & 0 & 0 \\ & & & c_{44} & 0 & c_{15} \\ & & & & c_{55} & c_{14} \\ & & & & & \tfrac{1}{2}(c_{11}-c_{12}) \end{matrix}$$

Trigonal; Classes 32, 3m, $\bar{3}2/m$

$$\begin{matrix} c_{11} & c_{12} & c_{13} & c_{14} & 0 & 0 \\ & c_{11} & c_{13} & -c_{14} & 0 & 0 \\ & & c_{33} & 0 & 0 & 0 \\ & & & c_{44} & 0 & 0 \\ & & & & c_{55} & c_{14} \\ & & & & & c_{66} \end{matrix}$$

Hexagonal System; Classes $\bar{6}$, 6, 6/m, $\bar{6}m2$, 6mm, 622, 6/m 2/m 2/m

$$\begin{matrix} c_{11} & c_{12} & c_{13} & 0 & 0 & 0 \\ & c_{11} & c_{13} & 0 & 0 & 0 \\ & & c_{33} & 0 & 0 & 0 \\ & & & c_{44} & 0 & 0 \\ & & & & c_{44} & 0 \\ & & & & & \tfrac{1}{2}(c_{11}-c_{12}) \end{matrix}$$

The same form of the matrices holds for the elastic compliances s_{ij}, with the following exceptions:

1. $\frac{1}{2}(c_{11} - c_{12})$ is replaced by $2(s_{11} - s_{12})$.
2. In the trigonal system, in matrix positions 45 and 56, c_{15} and c_{14} are replaced by $2s_{15}$ and s_{14}, respectively, in classes 3, $\bar{3}$. In position 56, c_{14} is replaced by $2s_{14}$ in the matrix for the remaining classes.

Because of the importance of isotropic materials in terms of material science or engineering applications, we give the terms which are used most frequently to describe their behavior. We start with two Lamé constants, λ and G, where G is the shear modulus and $\lambda + 2G$ the longitudinal modulus. Both, as we shall see later, may be calculated directly from measured parameters. c_{12} is replaced by λ and $\frac{1}{2}(c_{11} - c_{12})$ is replaced by G, so the matrix may be written as

$$\begin{pmatrix} (\lambda + 2G) & \lambda & \lambda & 0 & 0 & 0 \\ & (\lambda + 2G) & \lambda & 0 & 0 & 0 \\ & & (\lambda + 2G) & 0 & 0 & 0 \\ & & & G & 0 & 0 \\ & & & & G & 0 \\ & & & & & G \end{pmatrix}$$

All the important constants, for an isotropic solid, are expressible in terms of λ and G. They are given below.

Longitudinal modulus: $\lambda + 2G$

Young's modulus: $G \dfrac{3\lambda + 2G}{\lambda + G}$

Bulk modulus: $\lambda + \dfrac{2G}{3}$

Poisson's ratio: $\dfrac{\lambda}{2(\lambda + G)}$

2.4 Plane Wave Propagation

We now concern ourselves with the rate at which a stress-induced strain propagates through a solid. The reaction of a medium to a stress appears on the macroscopic scale of things to be instantaneous. If a bar is pushed at one end, the opposite end moves with no visible time delay as the bar translates. If the opposite end of the bar is fixed, the reaction force appears to arise instantaneously. Consider a medium so large that its inertia will resist any translation due to a small stress applied over a very small region

26   Elastic Constants and Their Measurement

of the surface. We consider then a semi-infinite body, and concern ourselves with the reaction which arises within the body in the neighborhood of the applied stress.

The force applied to the material will result in three localized strain components for which Hooke's law holds. Consider the component in the direction of the stress σ. Since the stress is small, Hooke's law applies; so the resulting strain will be proportional to the applied stress, and upon release of the stress there is elastic recovery. The resultant strain stresses the material immediately in its path so that a displacement (mechanical) wave is propagated through the material. If the stress is uniform the energy propagates as a plane wave; and because it travels through a semi-infinite medium, there are no perturbations due to boundary effects. In this section, we consider the equations of motion of such a mechanical wave and relate its velocity to the elastic constants.

Define the orthogonal reference axes X_1, X_2, X_3 as before. As a consequence of the elastic behavior, and using the fundamental definition of strain given by Eq. (2.1) and Fig. 2.2, we have Eqs. (2.7) expressing the stress components in terms of the six strain components (displacements), for an element of volume (in the matrix notation), explicitly as the six similar expressions in Eq. (2.12), the first and last being given below:

$$\sigma_1 = c_{11}\frac{\partial u_1}{\partial X_1} + c_{12}\frac{\partial u_2}{\partial X_2} + c_{13}\frac{\partial u_3}{\partial X_3} + c_{14}\left(\frac{\partial u_2}{\partial X_3} + \frac{\partial u_3}{\partial X_2}\right)$$

$$\vdots \qquad\qquad + c_{15}\left(\frac{\partial u_3}{\partial X_1} + \frac{\partial u_1}{\partial X_3}\right) + c_{16}\left(\frac{\partial u_2}{\partial X_1} + \frac{\partial u_1}{\partial X_2}\right) \quad (2.12a)$$

$$\sigma_6 = c_{61}\frac{\partial u_1}{\partial X_1} + c_{62}\frac{\partial u_2}{\partial X_2} + c_{63}\frac{\partial u_3}{\partial X_3} + c_{64}\left(\frac{\partial u_2}{\partial X_3} + \frac{\partial u_3}{\partial X_2}\right)$$

$$+ c_{65}\left(\frac{\partial u_3}{\partial X_1} + \frac{\partial u_1}{\partial X_3}\right) + c_{66}\left(\frac{\partial u_2}{\partial X_1} + \frac{\partial u_1}{\partial X_2}\right) \quad (2.12f)$$

u_1, u_2, and u_3 are the components of the displacement referred to the X_1, X_2, X_3 directions. Since the material is in dynamic equilibrium, Newton's second law of motion applies. Using tensor notation,

$$\frac{\partial \sigma_{11}}{\partial X_1} + \frac{\partial \sigma_{12}}{\partial X_2} + \frac{\partial \sigma_{13}}{\partial X_3} = \rho\frac{\partial^2 u_1}{\partial t^2} \quad (2.13a)$$

$$\frac{\partial \sigma_{21}}{\partial X_1} + \frac{\partial \sigma_{22}}{\partial X_2} + \frac{\partial \sigma_{23}}{\partial X_3} = \rho\frac{\partial^2 u_2}{\partial t^2} \quad (2.13b)$$

$$\frac{\partial \sigma_{31}}{\partial X_1} + \frac{\partial \sigma_{32}}{\partial X_2} + \frac{\partial \sigma_{33}}{\partial X_3} = \rho\frac{\partial^2 u_3}{\partial t^2} \quad (2.13c)$$

Note: We have differentiated the stress tensor $\partial(\sigma_{ij})/\partial X_j$ and since $\sigma_{ij} = \sigma_{ji}$, the left-hand side of (2.13) is also symmetric.

The set (2.13a to c) contains the equations of motion for the displacements u_1, u_2, u_3 resulting from their respective stress components.

For a periodic infinite plane wave, the displacements u_1, u_2, u_3 at any time t may be expressed as

$$u_1 = U_{10} \sin(lX_1 + mX_2 + nX_3 - vt) \tag{2.14a}$$

$$u_2 = U_{20} \sin(lX_1 + mX_2 + nX_3 - vt) \tag{2.14b}$$

$$u_3 = U_{30} \sin(lX_1 + mX_2 + nX_3 - vt) \tag{2.14c}$$

where $l, m, n =$ direction cosines of the wave normal with respect to the axes X_1, X_2, X_3 (see Fig. 2.3)

$v =$ phase velocity

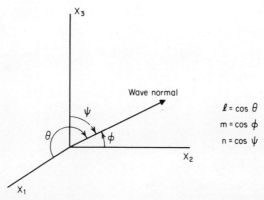

Fig. 2.3 Direction numbers and direction cosines, referred to reference axes X_1, X_2, X_3.

The nontrivial solutions of Eqs. (2.14) are those for which the amplitude is nonzero. This is met for all the amplitudes when the following condition is fulfilled:

$$\begin{bmatrix} (L_1 - \rho v^2)U_{10} & M_1 U_{20} & N_1 U_{30} \\ M_1 U_{10} & (M_2 - \rho v^2)U_{20} & N_2 U_{30} \\ N_1 U_{10} & N_2 U_{20} & (N_3 - \rho v^2)U_{30} \end{bmatrix} = 0 \tag{2.15}$$

The L, M, and N coefficients arise from the term-by-term differentiation of Eqs. (2.14a to c), and the determinant is formed by setting each equation equal to zero after collecting terms. The coefficients are listed below:

$$L_1 = c_{11}l^2 + c_{66}m^2 + c_{55}n^2 + 2c_{15}\,ln + 2c_{16}\,lm + 2c_{56}\,mn$$

$$M_1 = c_{16}l^2 + c_{26}m^2 + c_{45}n^2 + lm(c_{12} + c_{66}) + ln(c_{14} + c_{56}) + mn(c_{25} + c_{46})$$

$$N_1 = c_{15}l^2 + c_{46}m^2 + c_{35}n^2 + lm(c_{14} + c_{65}) + ln(c_{13} + c_{55}) + mn(c_{36} + c_{45})$$

$$M_2 = c_{66}l^2 + c_{22}m^2 + c_{44}n^2 + 2c_{26}\,lm + 2c_{46}\,ln + 2c_{44}\,mn$$

$$N_2 = c_{56}l^2 + c_{24}m^2 + c_{34}n^2 + lm(c_{25} + c_{46}) + ln(c_{36} + c_{45}) + mn(c_{34} + c_{44})$$

$$N_3 = c_{55}l^2 + c_{44}m^2 + c_{33}n^2 + 2c_{45}\,lm + 2c_{35}\,ln + 2c_{34}\,mn \tag{2.16}$$

The general solution of the determinant (2.15) (known as the secular equation) is one involving the velocity to the third power. There will, in general, be three different velocities associated with each solution of (2.15) for every choice of propagation direction given by the direction cosines l, m, and n. The most general solutions involving all the elastic constants (triclinic system) and any arbitrary direction add up to a most formidable task, which has been accomplished by Neighbors and Schacher.[6] However, an immediate reduction may be achieved if the propagation directions are simple (i.e., 100, 110, 111). For these directions, the coefficients defined by Eq. (2.16) reduce to

1. 100 direction

$$L_1 = c_{11}; \quad M_1 = c_{16}; \quad N_1 = c_{15}; \quad M_2 = c_{66}; \quad N_2 = c_{56}; \quad N_3 = c_{55} \quad (2.17a)$$

2. 110 direction

$$\begin{aligned} L_1 &= c_{11} + c_{66} + 2c_{16}; & M_1 &= c_{16} + c_{26} + c_{12} + c_{66} \\ N_1 &= c_{15} + c_{46} + c_{14} + c_{65}; & M_2 &= c_{66} + c_{22} + 2c_{26} \\ N_2 &= c_{56} + c_{24} + c_{25} + c_{46}; & N_3 &= c_{55} + c_{44} + 2c_{45} \end{aligned} \quad (2.17b)$$

3. 111 direction

$$\begin{aligned} L_1 &= c_{11} + c_{66} + c_{55} + 2(c_{15} + c_{16} + c_{56}) \\ M_1 &= c_{16} + c_{26} + c_{45} + c_{12} + c_{66} + c_{14} + c_{56} + c_{25} + c_{46} \\ N_1 &= c_{15} + c_{46} + c_{35} + c_{14} + c_{65} + c_{13} + c_{55} + c_{36} + c_{45} \\ M_2 &= c_{66} + c_{22} + 3c_{44} + 2(c_{26} + c_{46}) \\ N_2 &= c_{56} + c_{24} + c_{34} + c_{25} + c_{46} + c_{36} + c_{45} + c_{34} + c_{44} \\ N_3 &= c_{55} + c_{44} + c_{33} + 2(c_{45} + c_{35} + c_{34}) \end{aligned} \quad (2.17c)$$

If the principal axes are made to coincide with the three crystallographic axes and for minerals of the *cubic system*, the reduction of the elastic constants due to symmetry is imposed and the solutions for the secular equations are

1. 100

$$\rho v^2 = c_{11} \text{ (long.)}; \quad \rho v^2 = c_{44}, \quad \rho v^2 = c_{44} \text{ (shear)} \quad (2.18a)$$

2. 110

$$\begin{aligned} \rho v^2 &= c_{44} \text{ (shear)}; \quad \rho v^2 = \tfrac{1}{2}(c_{11} - c_{12}) \text{ (shear)} \\ \rho v^2 &= \tfrac{1}{2}(c_{11} + c_{12} + 2c_{44}) \text{ (long.)} \end{aligned} \quad (2.18b)$$

3. 111

$$\begin{aligned} \rho v^2 &= \tfrac{1}{3}(c_{11} - c_{12} + c_{44}) \text{ (shear)} \\ \rho v^2 &= \tfrac{1}{3}(c_{11} - c_{12} + c_{44}) \text{ (shear)} \\ \rho v^2 &= \tfrac{1}{3}(c_{11} + 2c_{12} + 4c_{44}) \text{ (long.)} \end{aligned} \quad (2.18c)$$

Since the cubic system has only three independent elastic constants, they may be determined explicitly from measurements of the wave velocities of shear and compressional modes, propagated in any two of these directions. For more complicated symmetry, the complexity of the solutions of the secular equation increases rapidly. These problems are considered in detail by several authors.[6-9] The problem of measurement in an arbitrary direction of a cubic solid is considered explicitly by Markham.[10]

If we apply the same reasoning to isotropic materials, we note that the secular equation in any principal direction is

$$\left\{ \begin{matrix} (c_{11} - \rho v^2) & 0 & 0 \\ & [\tfrac{1}{2}(c_{11} - c_{12}) - \rho v^2] & 0 \\ & & [\tfrac{1}{2}(c_{11} - c_{12}) - \rho v^2] \end{matrix} \right\} = 0$$

for which we have

$$\rho v^2 = c_{11} \text{ (long.)} \quad \text{and} \quad \rho v^2 = \tfrac{1}{2}(c_{11} - c_{12}) \text{ (shear)}$$

so that two measurements of velocity, one longitudinal and one shear, performed in any direction, define the elastic behavior of isotropic solids. It is fortunate that most polycrystalline materials exhibit an isotropic behavior (due to an essential lack of preferred orientation) because it permits use of the simplest of the measurement schemes in determining their elastic behavior.

2.5 Calculation of the Isotropic Bulk and Shear Moduli

If the individual grains of a homogeneous polycrystalline material are randomly oriented, the polycrystalline sample will behave as an isotropic elastic solid, for if a sufficient number of grains are included in any direction, all possible orientations will be included and the average property in any direction will be the same as in any other. We will call such solids, which include most of the technological materials (e.g., metals, ceramics, most plastics), quasi-isotropic, to distinguish them from those materials which are truly isotropic, such as glass. It should be kept in mind that many of the schemes employed in fabricating these materials induce a preferred orientation to a greater or lesser extent. The processes of rolling and extrusion are perhaps among the most obvious. Therefore, it is not safe to assume that because a material is polycrystalline, it will exhibit an isotropic behavior. It will exhibit such behavior only if the orientation of the constituent crystallites are randomly oriented. It is advisable to check isotropy by performing measurements along several directions.

The importance of polycrystalline materials heightens the desirability of calculating isotropic elastic properties from measurements upon single crystal specimens. Thus, one may compute the "aggregate" properties of materials and select the one material most suitable for the intended application.

The problem of computing the elastic properties of a quasi-isotropic solid from the measured elastic moduli of a single crystal is not simple. Two averaging schemes have been derived—one by Voigt[11] which is based upon the assumption that the stress is uniform everywhere within the sample. Application of this assumption leads to a set of equations expressing the shear and bulk moduli in terms of elastic stiffnesses. Alternately, one may assume that the strain is uniform throughout the sample, and that the stress varies from point to point within the body. This assumption, made by Reuss,[12] led to a different set of equations relating the reciprocals of the shear and bulk moduli to the single crystal elastic compliances. Hill[13] examined this question and on theoretical grounds demonstrated that the Voigt and Reuss averaging schemes determine the upper and lower limits, within which the values of the shear and bulk moduli, determined on a quasi-isotropic body of the same substance, must fall. Hill suggested that the arithmetic mean of the values calculated, using the Voigt and Reuss schemes, be used to express the most probable value for the quasi-isotropic solid. It was not until 1963 that the Voigt-Reuss-Hill approximation was tested and its validity demonstrated.[14] This scheme is now accepted as a means of computing the shear and bulk moduli for polycrystalline solids based upon the measured values of the single crystal elastic constants.

The general expressions for the Voigt and Reuss schemes are: for bulk modulus by the Voigt approximation (upper limit)

$$B_v = \tfrac{1}{9}(c_{11} + c_{22} + c_{33}) + \tfrac{2}{9}(c_{12} + c_{23} + c_{13}) \qquad (2.19)$$

and for shear modulus by the Voigt approximation (upper limit)

$$G_V = \tfrac{1}{15}(c_{11} + c_{22} + c_{33}) - \tfrac{1}{15}(c_{12} + c_{23} + c_{13}) + \tfrac{1}{5}(c_{44} + c_{55} + c_{66}) \qquad (2.20)$$

and for bulk modulus by the Reuss approximation (lower limit)

$$\frac{1}{B_R} = (s_{11} + s_{22} + s_{33}) + 2(s_{12} + s_{23} + s_{13}) \qquad (2.21)$$

and for shear modulus by the Reuss approximation (lower limit)

$$\frac{1}{G_R} = 4(s_{11} + s_{22} + s_{33}) - 4(s_{12} + s_{23} + s_{13}) + 3(s_{44} + s_{55} + s_{66}) \qquad (2.22)$$

Equations (2.19) to (2.22) may be cast into forms applicable to each crystal class by applying the appropriate elastic stiffness or compliance matrix. Thus for the cubic system, where

$$c_{11} = c_{22} = c_{33}, \; c_{12} = c_{23} = c_{13}, \; c_{44} = c_{55} = c_{66}$$

and

$$s_{11} = s_{22} = s_{23}, \; s_{12} = s_{23} = s_{13}, \; s_{44} = s_{55} = s_{66}$$

we have

$$B_V = \tfrac{1}{3}(c_{11} + c_{12})$$
$$G_V = \tfrac{1}{5}(c_{11} - c_{12} + 3c_{44}) \tag{2.23}$$

$$B_R = [3(s_{11} + 2s_{12})]^{-1}$$
$$G_R = \left[\frac{4(s_{11} - s_{12})}{5} + \frac{3s_{44}}{5}\right]^{-1} \tag{2.24}$$

Equations for all the remaining crystal classes may be obtained in a similar way, and using the equations given in Table 1.1, all the remaining isotropic elastic properties may be computed, including the propagation velocity of acoustic waves, if the density is also known.

2.6 Adiabatic Character of Acoustic Measurements

An aspect of elastic properties measurement that may be easily overlooked involves the adiabatic character of values derived from measurements using acoustic methods. This in turn is connected to the conditions under which work is done in deforming the specimen. In static methods the specimen is squeezed, pulled, or twisted slowly, and the resultant deformation is measured. During this process, which occurs slowly relative to the rate of heat flow, the specimen is firmly and intimately in contact with a large apparatus functioning as a heat sink. Static measurements consequently amount to isothermal ones. By contrast, when acoustic techniques are employed, the transit time of a compressional or extensional part of the wave passing through a region of the sample is very rapid compared to the rate of heat flow. There is no transfer of thermal energy between the region upon which work is being done in the specimen and its surroundings. Consequently the process is adiabatic and, for processes which involve extensions or compressions, yields different values for the elastic constants from those measured isothermally (i.e., by static methods). Deformations which involve pure shear only will yield the same result by either isothermal or adiabatic methods. To see why this is so in a qualitative fashion, compare the measurements performed under isothermal and adiabatic conditions to a given amount of work done in straining the sample. In the isothermal system, part of the energy is converted to heat by dissipative forces, and such heat is removed to the surroundings. The remaining energy is stored in the elastic deformation, as the sample undergoes an isothermal strain ε_i. In the adiabatic case, the heat goes to raise the temperature of, say, the region under compression. This causes a local rise in temperature together with a local expansion. The expansion (or contraction in the case of a dilatation) is opposite to the dimensional change (strain) induced by the stress. Hence, the isothermal strain ε_i is greater than the adiabatic strain ε_a, and for the same applied stress, the material appears *stiffer* under adiabatic conditions than it does under isothermal conditions. Since in a pure shear there is no volume change, no work is done and the "isothermal" and "adiabatic" moduli or compliances are equal.

The conversion between the isothermal and adiabatic moduli is through thermodynamics. To compute the one from measurements of the other requires a relationship involving the work done in expanding or contracting the sample in the adiabatic case. To treat this quantitativity, consider the following effects. (For simplicity we will consider the thermal expansivity α to be a scalar, which is valid for cubic material. For lower symmetry, it is a tensor and appropriate subscripting is required to express the effect of the direction along which measurements are made.) For a given amount of work, the change in strain will be the sum of the elastic and thermal effects, and writing that change as a total differential,

$$d\varepsilon_i = \left(\frac{\partial \varepsilon_i}{\partial \sigma_j}\right)_T d\sigma_i - \left(\frac{\partial \varepsilon_i}{\partial T}\right)_\sigma dT \qquad (2.25a)$$

where we recognize that

$$\left(\frac{\partial \varepsilon_i}{\partial \sigma_j}\right)_T = s_{ij}^T \qquad \left(\frac{\partial \varepsilon_i}{\partial T}\right)_\sigma = \alpha \qquad (2.25b)$$

where the superscript T indicates the isothermal process (a superscript s will denote an adiabatic process).

In addition, we also consider the entropy change as a result of a stress and temperature (where S is the entropy),

$$dS = \left(\frac{\partial S}{\partial \sigma_j}\right)_T d\sigma_j + \left(\frac{\partial S}{\partial T}\right)_\sigma dT \qquad (2.26)$$

where $(\partial S/\partial \sigma_j)_T =$ "piezocaloric" coefficient
$(\partial S/\partial T)_\sigma =$ heat capacity at constant pressure C_p

Now, for a reversible process, we can write the Gibbs free energy as

$$dG = l\, d\sigma_j - S\, dT \qquad (2.27)$$

where the first term on the right is the PV term. Since Eq. (2.27) is an exact differential,

$$\left(\frac{\partial l}{\partial T}\right)_\sigma = -\left(\frac{\partial S}{\partial \sigma_j}\right)_T \tag{2.27a}$$

so that we have $-(\partial S/\partial \sigma)_T = \alpha l$ (the thermal expansivity times the length), we can express Eqs. (2.25) and (2.26) now as

$$\begin{aligned} d\varepsilon_i &= s_{ij}^T d\sigma_j - \alpha\, dT \\ dS &= -\alpha\, d\sigma_j + C_p\, dT \end{aligned} \tag{2.28}$$

For the adiabatic condition, $dS = 0$; so we have

$$dT = \frac{\alpha l T}{C_p} d\sigma_j$$

and substituting this into upper Eq. (2.28)

$$d\varepsilon_i = s_{ij}^T d\sigma_j - \frac{\alpha l T}{C_p} d\sigma_j$$

valid at constant entropy. Then, dividing by $d\sigma_j$

$$\left(\frac{\partial \varepsilon_i}{\partial \sigma_j}\right)_s = s_{ij}^s = s_{ij}^T - \frac{\alpha^2 l T}{C_p}$$

or

$$s_{ij}^T = s_{ij}^s \left(1 + \frac{\alpha^2 l T}{C_p s_{ij}^s}\right) \tag{2.29}$$

which quantitatively relates compliances measured under adiabatic and isothermal conditions. The adiabatic compliance is smaller than the isothermal compliance, which again emphasizes the fact that under an adiabatic process, the material is stiffer.

Because the compressibility is a purely thermodynamic quantity, it is convenient to express Eq. (2.29) explicitly in this case as

$$\chi_T = \chi_S \left(1 + \frac{\alpha^2 V T}{C_p \chi_S}\right) \tag{2.30a}$$

or for the bulk modulus

$$B_T = B_S \left(1 + \frac{\alpha^2 V T B_s}{C_p}\right)^{-1} \tag{2.30b}$$

Using Eqs. (2.30a or b) one can compute the isothermal compressibility or bulk modulus from acoustically determined values, and this together with the shear modulus is sufficient to compute the remaining isothermal moduli for isotropic materials, using the relations given in Table 1.1.

These relations play a significant role in evaluating the quality of data, for they make possible a direct comparison of moduli determined by static and dynamic methods.

REFERENCES

1. Huntington, H. B.: The Elastic Constants of Crystals, in Seitz and Turnbull (eds.), "Solid State Physics," vol. 7, Academic, New York, 1958.
2. Nye, J. F.: "Physical Properties of Crystals," Oxford University Press, New York, 1960.
3. Love, A. E. H.: "A Treatise on the Mathematical Theory of Elasticity," 4th ed., Dover, New York, 1944.
4. Mason, W. P.: "Physical Acoustics and the Properties of Solids," Van Nostrand, Princeton, N.J., 1958.
5. Fumi, F. G.: Physical Properties of Crystals: The Direct Inspection Method, *Acta Crystallogr.*, **5**:44 (1952).
6. Neighbors, J. R., and G. E. Schacher: Determination of Elastic Constants from Sound-Velocity Measurements in Crystals of General Symmetry, *J. Appl. Phys.*, **38**:5366 (1967).
7. Neighbors, J. R.: An Approximation for the Determination of the Elastic Constants of Single Crystals, *J. Acoust. Soc. Am.*, **26**:865 (1954).
8. Borgnis, F. E.: Specific Directions of Longitudinal Wave Propagation in Anisotropic Media, *Phys. Rev.*, **98**:1000 (1955).
9. McSkimin, H. J.: Ultrasonic Methods for Measuring the Mechanical Properties of Liquids and Solids, in W. P. Mason (ed.), "Physical Acoustics," vol. 1, pt. A, Academic, New York, 1965.
10. Markham, M. F.: Measurement of Elastic Constants by the Ultrasonic Pulse Method, *Brit. J. Appl. Phys.*, Suppl. no. 6, S56–S63 (1967).
11. Voigt, W.: "Lehrbuch der Kristallphysik," 739 pp., Teubner, Leipzig, 1928.
12. Reuss, A.: Berechung der Fliessgrenze von Mischkristallen auf Grund der Plastizitätsbedingung für Einkristalle, *Z. Angew. Math. Mech.*, **9**:49 (1929).
13. Hill, R.: Elastic Behaviour of a Crystalline Aggregate, *Proc. Phys. Soc. London*, **65A**:350 (1952).
14. Anderson, O. L.: A Simplified Method for Calculating the Debye Temperature, *J. Phys. Chem. Soc.*, **24**:909 (1963).

CHAPTER THREE

The Determination of Velocity of Propagation

3.1 Introduction

While it would appear, at first glance, that the determination of the velocity of propagation should be a simple matter, in fact, obtaining it from the direct measurement of the specimen length and the travel time of an acoustic wave or impulse is only a relatively recent achievement. The difficulty has lain principally in the precise measurement of very small time intervals. For example, in a typical ceramic material, the longitudinal velocity may be about 9 km/s[1-3] and a useful specimen length is about 5 cm. The travel time, or the time for a plane wave to propagate from one end of the specimen to the other, is 5.5 μs. Hence, for a precision of 0.1 percent, a resolution of at least 5 ns is required. The ability to obtain this time resolution is further complicated by the fact that the signal-to-noise ratio is small, and the signal quality is degenerated by the dispersive nature of the specimen.

The capability of measuring these time-delayed signals has therefore been closely related to achievements in electronic component and circuit designs. The circuitry capable of responding to extremely fast rise times with high counting rates, broad but flat frequency responses, and high sensitivity was rapidly adapted as components in apparatus for determining sound velocity propagation. It is now possible to purchase off-the-shelf components and assemble the most sophisticated apparatus capable of great accuracy, while only a short time ago, the experimenter interested in determining the elastic properties of materials dynamically also had to have an expert knowledge of the electronic arts.

In the sections which follow, we shall develop the various techniques in current use from an historical viewpoint. This has the advantage of leading from the simpler, more direct methods to the more sophisticated, accurate, but also more complex techniques. We shall consider the methods as falling into three categories: (1) transit time measurements, (2) pulse-echo methods, (3) ultrasonic interferometry. No attempt at completeness in surveying the literature on this subject is attempted. Excellent reviews have been written by McSkimin[4] and by Anderson and Liebermann.[5] We will concern ourselves here with detailed discussions concerning the application of several methods in each of the above-mentioned categories.

3.2 Transit-Time Measurements

The earlier work based upon the direct measurement of the transit times followed on the heels of radar developments. Indeed, in this early work, radar-ranging scopes, timing circuits, and pulse generators were used. A block diagram of the arrangement employed by Hughes et al.[6] is shown in Fig. 3.1.

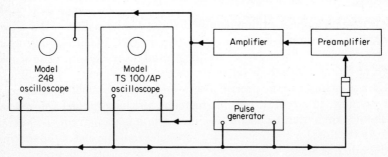

Fig. 3.1 Block diagram of the apparatus used by Hughes et al.

The basic timing circuit employed was the TS100/AP oscilloscope, with a circular sweep of 12.836 μs/revolution, derived from an 80.86-kHz quartz crystal used to provide a trigger for the pulse generator. The sweep of the Model 248 oscilloscope was also triggered by the TS100/AP unit. Pulses were applied to X-cut (longitudinal mode) quartz crystals bonded to one end of the specimen. The signal, delayed by its passage through the specimen, is received by a similar transducer bonded at the opposite end; the output of this receiving transducer is displayed on the oscilloscopes, and the arrival time is measured. With this arrangement, a precision of 0.03 μs was achieved, or about 1 percent on the basis of our earlier example.

The arrival time of the delayed signal is determined from its observed first motion, and the accuracy of this and similar methods is largely limited by the ability to detect this first motion as it rises out of the noise. This measured time delay (t) of the specimen (or more exactly the delay in the specimen, bond, transducer, and electronic components) is used to calculate the velocity of a *longitudinal* wave. This is *not* the bar velocity associated with Young's modulus, but rather that linked with the Lamé constants (λ and G) for an isotropic medium; it is related to the bulk modulus and shear modulus by

$$v_l = \left[\frac{1}{\rho}\left(B + \frac{4}{3}G\right)\right]^{1/2} \tag{3.1}$$

where $B =$ bulk modulus
$G =$ shear modulus
$\rho =$ density

Determination of the shear velocity would establish G, and thereby all the other elastic constants of the isotropic solid are defined by the relations given in Table 1.1.

The velocity of the first arrival of longitudinal waves was found to be independent of both rod length and diameter. However, a set of subsequent arrivals were observed to be independent of length but related to the specimen diameter. These arrivals resulted from the mode conversion of longitudinal to shear, and back to longitudinal waves which occurred along the specimen boundary. The additional delay Δt, introduced by each set of mode conversions, is given by

$$\Delta t = d \left(\frac{v_l^2 - v_s^2}{v_l v_s}\right)^{1/2} \tag{3.2}$$

From the observation of the time of arrival of both the direct dilatational wave and mode-converted wave, it is possible to determine the velocities of both the longitudinal and shear waves.

For many materials, this effect is obscured because of the dispersive nature of the medium. However, it has been successfully applied to the simultaneous measurement of both longitudinal and shear waves[7-9] in rocks at ambient conditions, and also as a function of temperature, for the determination of the elastic properties of iron, aluminum, and fused quartz;[10] and the method has also been extended to anisotropic media by Araki,[11,12] who has measured the properties of single crystals of the cubic system.

Errors in this method of measurement occur if the rise time of the applied pulse is low, or if the sample strongly attenuates the high frequencies. In either case, the magnitude of first motion of the received signal is obscured. If the longitudinal wave signal is dispersed, it will obscure the secondary arrivals and a separate determination of the shear velocity is required. This is readily accomplished by substituting Y-cut or AC-cut* quartz shear transducers for the driving and receiving transducers. In contrast to other dynamic techniques, the advantage of these pulse transmission methods is that the signal can usually be propagated through very lossy materials, such as porous ceramics.

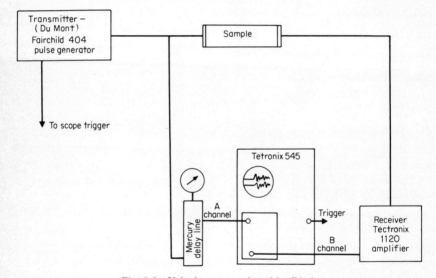

Fig. 3.2 Velocimeter employed by Birch.

Birch[13] has modified the technique described by employing a calibrated variable mercury delay line in order to determine the travel time of the pulse through the specimen. The arrangement is shown in Fig. 3.2. Velocity is determined in the following way: A pulse, at a low repetition rate (300 to 1,000 Hz) from the pulse generator, is applied simultaneously to sending transducers (e.g., X-cut quartz) attached to the sample and to the mercury delay line. The low repetition rate is dictated by the requirement

* The notation follows the Standards on Piezoelectric Crystals, *Proc. IRE*, 1378–1395 (1949).

that the wave train following a received pulse attenuates completely before the signal is reapplied. A triggering signal from the generator synchronizes the horizontal sweep of the oscilloscope with the pulse generator. Because the mercury does not attenuate the signal beyond the sensitivity of the vertical amplifiers of the oscilloscope, the output of the receiving transducer is applied directly to the A channel of the dual-trace plug-in unit. The signal through the sample may be highly attenuated, and, frequently, preamplification is necessary before applying it to the B channel of the vertical amplifiers. The time base of the oscilloscope is adjusted so that the applied and received signals are both visible on the B trace. The two base lines are superimposed, and the length of the mercury column adjusted until the first motion of the incoming wave train from the mercury column is aligned exactly over the initial rise of the applied pulse. The length of the column is now increased until the signal from the mercury column is aligned over the first motion of the wave train arriving from the specimen. The difference between the final and initial lengths (Δl) of the mercury column (as determined with a micrometer to within 0.001 in) provides the basis for the calculation of the velocity. The transit time for the specimen-transducer assembly is the product of this length difference and the calibrated velocity of the mercury column v_{Hg}. The velocity is given simply by

$$v = \frac{l v_{\mathrm{Hg}}}{\Delta l} \tag{3.3}$$

where $l =$ sample length.

The mercury line may be calibrated against a standard time mark generator to determine v_{Hg}.

With care, a precision of about 0.1 percent can be achieved, and a resolution of about 20 ns obtained. Such simple devices as maintaining sharp oscilloscope traces, achieved through adjustment of the focus and astigmatism controls, and varying signal levels so that the shape of the rise times are similar help in obtaining maximum precision. The accuracy is not well determined, since the signal does not rise out of the noise with an infinite slope but starts at zero slope and increases rapidly but concavely upward. Hence the exact point in time when the first motion is initiated cannot be established unambiguously. This factor is one of the principal limitations upon accuracy. A further limitation arises out of the time delay introduced by the couplant between specimen and transducers and the transducers themselves. (These are very small, roughly 10 to 20 ns, and, for a specimen with a 10-μs delay, would introduce errors smaller than the precision of the method.) Finally, the errors in measuring specimen length and those involved in calibration of the mercury line further decrease the accuracy of the method.

Recently, Mattaboni and Schreiber[14] described a method for performing pulse transmission measurements with improvement in both precision and accuracy. The arrangement of equipment is shown in the block diagram of Fig. 3.3. The output of a stable, variable frequency oscillator (VFO), after being suitably divided, is used to trigger a Hewlett-Packard 214A (or other suitable) pulse generator. The output trigger of the pulse generator is used to synchronize the horizontal sweep of a Tektronix 535A

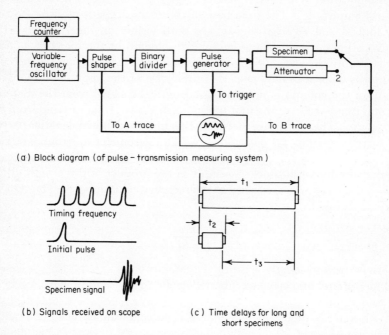

Fig. 3.3 (a) Block diagram of modified pulse-transmission method. (b) Schematic view of signals observed on oscilloscope. (c) Illustration of time delays when using combination of long and short specimen.

oscilloscope equipped with a Type AC dual-trace plug-in unit. The pulse generator output consists of a pulse with a repetition rate which is a known submultiple of the frequency of the VFO controlling it. The VFO frequency is measured with the frequency counter, which has an accuracy of 1 part in 10^8, so the instrumentation error in measuring the period (or time delay) is negligibly small. The pulses are applied through a manually operated switch (90-dB isolation) alternately either to a transducer bonded to one end of the specimen, or directly to the oscilloscope, where after suitable attenuation, it is displayed on, say, the B trace. The first arrival of the impulse transmitted through the specimen is received at the opposite end by a similar transducer and is displayed on the same oscilloscope trace.

The signal from the VFO is also fed to pulse-shaping circuits and displayed on the second trace (A trace, say) of the oscilloscope. This provides a continuously variable measure of the time delay with an accuracy equal to that of the frequency counter. In principle, the time interval between the initial pulse and received signal is determined by setting the VFO so that the period of one wavelength is an exact submultiple of the time delay to be measured.

The measurement of the time delay is performed according to the following steps:

1. Display the received signal from the specimen on the B trace of the oscilloscope (switch in position 1). With the aid of the pulse-shaping circuits and the Y amplifier of the oscilloscope, adjust the shape of the initial rise of the timing frequency (displayed on the A trace) to conform with the onset of motion of the received signal. Align the two by superimposing the traces and adjusting the VFO frequency so that the signals overlap exactly.

2. Display the initial pulse on the B trace (switch in position 2) by shifting the horizontal trace to earlier time. Count the number of cycles of timing frequency between the initial pulse and the delayed arrival from the specimen. This is the multiplicity M.

3. Using the attenuator, adjust the initial pulse to conform with the shape of the timing frequency and superpose the two traces by adjusting the pulse position control of the pulse generator.

The sequence of steps is repeated until setting the VFO does not perturb the coincidence of the timing frequency and the initial pulse. Under these conditions, the time delay of the specimen is given by

$$t = \frac{M}{F} - t' \qquad (3.4)$$

where t = actual specimen delay
F = frequency of the VFO
M = multiplicity (defined above)
t' = fixed delays arising principally as a result of end effects from the transducer-bond-specimen assemblies

Since the compared signals traverse the same path, delays introduced by the electronic equipment are made negligible. The magnitude of t' is small and will depend upon the mechanical impedance of specimen and transducer.

To test this technique, measurements were made upon a specimen of corundum (Lucalox) obtained from the General Electric Company. This material was chosen to evaluate the method because of its uniformity and availability, and also because its quality allowed cross-checking by other techniques. Measurements were performed for both longitudinal and shear wave velocities using both long and short specimens with the method described above. In addition, the time delay was determined for a small specimen, using the technique of pulse superposition with a frequency of 40 MHz (described in Sec. 3.4.2) and for the long specimen using the technique of bar resonance (see Chap. 4) employing frequencies lower than 30 kHz. The results indicate that the velocity is independent of frequency for this specimen. The precision of the pulse transmission measurements using the short specimen was 0.1 percent, and for the long specimen, the precision obtained was 0.05 percent. This represents the precision of several independent sets of measurements; for a given set of measurements the repeatability (i.e., the ability to match the reference and specimen first motion) was 0.025 percent.

The accuracy of the measurements is illustrated in Table 3.1, where the results obtained by the three different methods are compared. There is good agreement between the different techniques.

It was noted previously that a small systematic error (t') occurs due to the end effects. As can be seen in Table 3.1 this produces no serious error

TABLE 3.1 Comparison of Results of Measurements Obtained upon a Specimen of Al_2O_3

Method	Vibration	Specimen length, cm	Measured time delay, μs	Velocity, km/s	Time delay, μs/cm
Bar resonance	Longitudinal	20.35	18.802	10.823	0.9239
Pulse transmission:					
A	Longitudinal	20.35	18.802	10.823	0.9239
B	Longitudinal	1.2359	1.165	10.76	0.9239
Pulse superposition	Longitudinal	1.2359	1.1574	10.833	0.9230
Bar resonance	Shear	20.35	31.951	6.369	1.570
Pulse transmission:					
A	Shear	20.35	31.955	6.368	1.570
B	Shear	1.2359	1.988	6.307	1.584
Pulse superposition	Shear	1.2359	1.675	6.367	1.569

when measurements are performed upon a long specimen but represents an error of about 1 percent in the time delay of a short specimen. A frequently used method of dealing with the errors introduced by such fixed delays is to perform measurements upon a series of specimens of different lengths and extrapolating the delay time to zero length, thereby eliminating this systematic error.

An alternate arrangement utilizes a short length of the specimen, say 1 cm, for comparison purposes, where the short specimen replaces the direct path (switch position 2). Similar transducers are fixed to the ends of the small specimen, and since it has the same acoustic impedance, the end effects introduced are the same as those of the longer specimen and can be subtracted out. In this instance, for the short and long specimens (subscripts 2 and 1, respectively) the delay times are

$$t_1 = \frac{M_1}{F_1} - t'$$

$$t_2 = \frac{M_2}{F_2} - t' \qquad (3.5)$$

$$t_3 = \frac{M_1}{F_1} - \frac{M_2}{F_2}$$

The important point is that the error t' has been eliminated (Fig. 3.3c). In practice, the onset of motion of the signal delayed by the short specimen is used as the reference time in place of the initial pulse from the pulse generator. The path length for computing the velocity is then the difference in length of the two specimens. Results obtained using this method yielded values of $v_p = 10.830$ and $v_s = 6.367$ km/s in substantial agreement with the results obtained by pulse superposition.

Pulse-transmission methods have been used successfully on a variety of materials, and have the inherent advantage that they are applicable to lossy specimens. The technique is simple, and only a minimum amount of time need be expended in preparing the end surfaces of the specimen.

In Fig. 3.4 we see the response at the receiving transducer to the arrival of a longitudinal wave in a specimen of spinel. The long wave train following the initial motion and sweep of the trace is a result of several factors. The rise time of the input pulse is rich in higher frequencies (the steeper the slope, the higher the frequencies contained), and these are also transmitted through the specimen. If the specimen is dispersive, the frequencies will be sorted out according to their velocities in their passage through the specimen and the first motion will correspond to that frequency for which energy is propagated fastest. The frequencies arriving later will interfere with each other and contribute to the shape of the wave train. Energy arriving after the first motion, delayed by mode conversion at the specimen boundaries or by multiple reflections from the ends of the specimen, and energy scattered by diffraction effects will also contribute to the shape and nature of the wave train. In general, it is recommended that the ratio of specimen length to diameter be less than 10, and preferably less than 5 to minimize the amount of energy lost from the main signal by mode conversion and by reflection. These effects become more serious at lower frequencies.

It can be shown[15] that the shear wave propagates through a cylindrical specimen at a velocity which is independent of the diameter. However, this is not so for the longitudinal wave. The velocity at which this type of wave propagates through the specimen is highly dependent upon the boundary condition. This can be seen, in a general way, from the following considerations. A compressional wave propagating through a long,

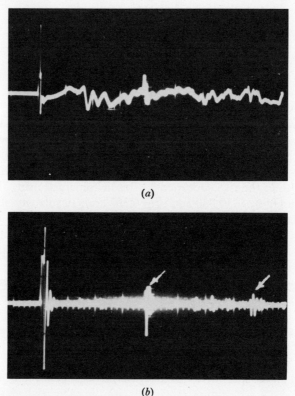

Fig. 3.4 Oscilloscope displays (*a*) of complex wave train following first arrival and (*b*) echoes seen after filtering the wave train.

thin rod represents uniaxial compressions and extensions where the displacements are related to the stresses through Young's modulus. However, in the bulk material, a uniaxial extension or compression results in displacements arising in the plane normal to the direction of propagation, which are expressed by Poisson's ratio. Hence, in the bulk, energy is coupled laterally during the passage of an extensional (compressional) wave motion. In the limit, as the rod is made thinner, it approaches a single chain of atoms, and there is no lateral cross-coupling. For thin rods and bars, the wave is propagated as the bar velocity given by

$$v_{\text{bar}} = \left(\frac{E}{\rho}\right)^{1/2} \tag{3.6}$$

where $E =$ Young's modulus
$\rho =$ density

For a thick rod, the velocity at which an extensional wave propagates through a bounded medium is a function of the ratio of diameter to wavelength (Fig. 3.5). Bancroft[16] has shown that as the ratio of diameter (a) to

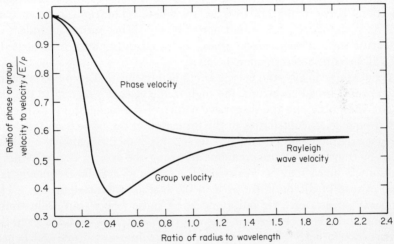

Fig. 3.5 Phase and group velocities of an extensional wave in a cylindrical rod. (*Adapted from "Physical Acoustics and the Properties of Solids," by W. P. Mason, © 1958 by Litton Educational Publishing, Inc., and reprinted by permission of Van Nostrand Reinhold Company, New York.*)

wavelength (λ) is increased, the wave propagates with the group velocity and approaches the Rayleigh (surface) wave velocity for thick rods (in terms of a/λ, a thin rod corresponds to $a/\lambda < 0.1$). However, in bulk samples the extensional wave is found to travel with the velocity given by the solution of the equation of motion for a longitudinal wave,

$$v_l = \left(\frac{\lambda + 2G}{\rho}\right)^{1/2} \tag{3.7}$$

Thus, by keeping the diameter of the specimen large in comparison to the *wavelength*, the longitudinal velocity rather than the bar velocity may be measured. For typical velocities (say between 5 and 10 km/s), and frequencies of about 10 MHz, the wavelengths are of the order of 100 μm. It is clear from Fig. 3.5 that a diameter in excess of 0.5 cm would satisfy this condition.

3.3 Pulse-Echo Methods

The wave train following the initial motion and first rise, as shown in Fig. 3.4a, is the result of the interference of waves arising from the combined effects of dispersion, reflection, and diffraction. In Fig. 3.4b, we see the same signal after it has been passed through a 5-MHz filter. With the aid of the filter, other frequencies in the wave train are suppressed, and we view the response of a narrow band of frequencies centered at about the filter frequency. Measurement of the time (or distance) between consecutive pairs of the arrivals shows that they are each separated by twice the arrival time of the first pulse. These are the echoes which arise from reflections between the ends of the specimen. The time between consecutive pairs of echoes is found not to be equal, but rather to undergo a systematic shift in their arrival times. This effect has been investigated[17] and will be considered further, below.

As noted previously, the energy contained in the applied pulse is distributed over a broad spectrum, the high frequencies are related to the rise time of the pulse, and the dispersive nature of the material, in sorting out these frequencies, contributes to the formation of the wave train following the first arrival. The potential of increasing both the accuracy and precision of transit-time measurements by using the echoes derived from a transmitted pulse may be realized by employing a radio-frequency (rf) pulse instead of the dc pulse used in the method discussed so far. The frequency spectra of a dc pulse and an rf pulse (the rf pulse illustrated is a tone burst) are compared in Fig. 3.6. The latter is obtained from the Fourier transforms. For the dc pulse, the maximum amplitude (and hence, energy content) is centered at low frequencies, the tail of the distribution extending toward the high frequencies, and the content of the latter in the distribution being a function of the rise time. For the tone burst, the energy content is both centered about the carrier frequency and narrower in range (Fig. 3.6b). This situation is somewhat idealized in that the shape of an rf pulse is generally not that of the tone burst (a consequence of the electronic circuitry responses), and, as a result, the spectrum is less compressed than indicated and skewed toward lower frequency. The longer the pulse width, the narrower the spectrum, and for a continuous tone, it would be a single frequency (Fig. 3.6c). Hence, for a highly dispersive material, longer pulse widths are recommended in order to increase signal purity and thereby help reduce or eliminate the complex wave train seen in Fig. 3.4. The acoustic signal derived from the applied pulse is degenerated further (in spectral purity and skewness), so that it is desirable to have the electronic signal initially be of as high quality as possible. The shape of the acoustic pulse derived from the electronic pulse has been studied by Redwood.[18] The further effect of the acoustic

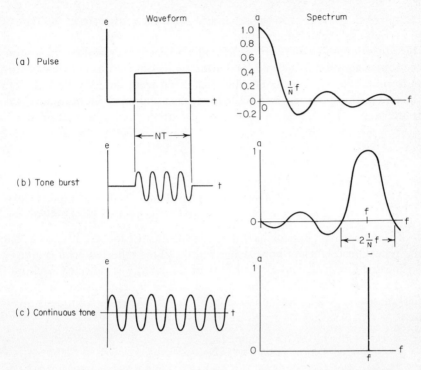

Fig. 3.6 Comparison of waveforms and spectrum envelopes for pulse, tone burst, and continuous tone signal.

coupling material upon the frequency distribution has been reported by Mason and McSkimin.[19]

From the foregoing, the advantages of an rf pulse in the determination of transit times between successive echoes are apparent, and the method was first used in this country by Pellam and Galt[20] in measurements upon liquids. The precision of pulse-echo measurements for small specimens is about an order of magnitude better than that for pulse-transmission measurements, and the sensitivity of the method permits the determination of the change in elastic constants of small (about 1-cm-long) specimens. This method has been used widely in determining the elastic behavior of solids as a function of pressure and temperature in studies concerned with the nature of the bonding forces and the equation of state of solids.[21-30]

The accuracy of the method suffers from the same difficulty encountered in pulse-transmission methods. An accurate determination of the time delay depends upon the ability to detect the first motion of the pulse arising from successive echoes, and it is made more uncertain by a distortion of the leading edge of each pulse that occurs upon every reflection.[17] This effect is illustrated in Fig. 3.7, where the initial voltage pulse is shown at

48   Elastic Constants and Their Measurement

the top of the figure. On the second through the fourth traces the type of distortion each echo suffers upon reflection may be seen. The nature of this distortion is evident in Figs. 3.7 and 3.8, where the problem concerning the initiating pulse is ignored, and attention is directed to the changes the pulse undergoes. The initial pulse has been transmitted, reflected from the end of the specimen, and returned to that end to which the transducer is attached. The reflection coefficient for stress amplitude is given by

$$R = \frac{Z_T - Z_S}{Z_T + Z_S} \quad (3.8)$$

where Z_T = acoustic impedance of the transducer
Z_S = acoustic impedance of the specimen (acoustic impedance is ρv, where ρ is the density and v the velocity in the medium)

If Z_A, the acoustic impedance for air, is substituted for Z_T, it is seen that the reflection coefficient approaches unity. This reflected echo impinges on the sending face, and a small part of the energy is reflected back into the specimen (Fig. 3.7A), while the remainder is transmitted through to the transducer. The transmission coefficient for stress amplitude is given by

$$T = \frac{2 Z_T}{Z_T + Z_S} \quad (3.9)$$

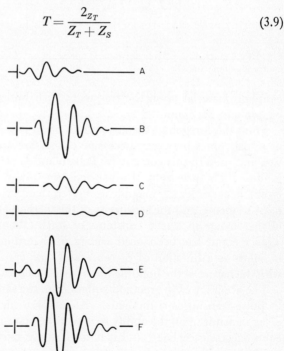

Fig. 3.7 Partial reflections of initial pulse at specimen-transducer boundaries (*A-E*). Wave passing into transducer is *F*.

and the energy passed into the transducer is reflected from the free end of the quartz and transmitted back into the specimen (Fig. 3.7B) delayed by one wavelength, since the resonant quartz crystal is about $\frac{1}{2}$-wavelength thick. In Figs. 3.7C and 3.7D there are additional components representing the portion of the energy which has been reflected back into the transducer from the loaded end, re-reflected from the free end, and transmitted back into the specimen (i.e., the decaying multiple reflections within the transducer). The wave which is returned to the specimen (Fig. 3.7E) propagates to the free end of the specimen and reflects back to form the second echo and is obtained by the summation of Figs. 3.7A to 3.7D. Figure 3.7F (and 3.8A) is the stress wave which passes through the quartz as the first

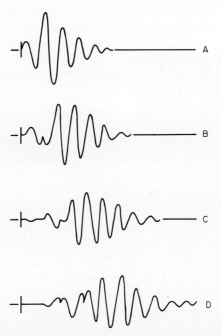

Fig. 3-8 Consecutive stress pulses in the transducer, each constructed as a sum of component terms (A) first, (B) second, (C) third, and (D) fourth echoes.

echo. Applying the same analysis to the second specimen return (having the shape shown in Figs. 3.7E and 3.8B), the interferences at the transducer-specimen interface result in a third pulse transmitted from the transducer with the shape shown in Fig. 3.8C. Similarly, a fourth echo pulse transmitted from the transducer to the specimen will have the shapes shown in Fig. 3.8D. The effect of the superposition of these multiple reflections, as noted by Eros and Reitz, are the following:

1. The first part of the echo pulse is strongly attenuated, and for the nth echo is proportional to R^n.

2. Cycles are added to the rear following each interfacial reflection to form an attenuating tail on the pulse.

3. The cycles forming the tail grow at the expense of those at the leading edge.

The effect of these phase shifts is to cause an added apparent delay in the transit time as measured between successive echoes. It is this effect that produced the departure from equality of the delay times measured between echoes seen in Fig. 3.4. As a consequence of this analysis, it is clear that measurements based upon the position of the observed first motion of subsequent echoes must contain this transit-time error. To eliminate the error, it is desirable to select a particular position within the pulse that is free of the effects introduced at the specimen-transducer interface. A useful procedure that has been adopted is to choose a cycle within the pulse (say the fifth) which is not modified by the distortion imposed on the preceding echoes.

A simple block diagram for the measurement of transit time by pulse-echo methods is given in Fig. 3.9.

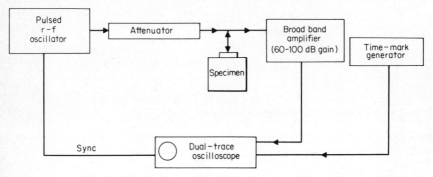

Fig. 3.9 Block diagram for the basic pulse-echo method.

In practice, the transducer is bonded to the specimen with wax, epoxy resin, or other suitable material. (These will be taken up in a later section.) The carrier frequency of the pulsed rf oscillator is set to the transducer resonance (10 MHz has been frequently used, but both higher and lower frequencies may be employed). The repetition rate of the oscillator is low (between 300 and 1,500 Hz) so that the echo train will have completely attenuated before the oscillator is pulsed again. It is advantageous to have the repetition frequency as high as possible without causing interference between echoes and applied pulse in order to intensify the image on the oscilloscope. In this operation, the transducer transmits the signal applied from the oscillator and functions as a receiver of the return echoes in the interval between pulses.

The measurement of the *round trip* time delay between echoes may be accomplished in a number of ways. A time-mark generator may be employed, as shown in the block diagram, to calibrate the delay circuits of the oscilloscope, which are then used to measure the transit time (A delayed by B time-base arrangement of the Tektronix 535A oscilloscope); or a mercury delay line may be used for transit-time measurement, as employed by McSkimin.[31] In this case, the precision is about 0.1 percent. The use of a dual-trace plug-in unit is very helpful, since the timing pulses can be displayed on one trace and the echo pattern displayed on the other. Figure 3.10a shows the echo pattern obtained from a specimen of spinel. Note the use of the unrectified echo trace in the expanded display of Fig. 3-10b. The arrows point to a cycle in the echoes which may be used in the measurement.

An alternate method, which uses a frequency counter to measure the period of the delay between echoes, has been described by May.[32] Papadakis[33] has favored this technique in his studies of velocity and attenuation behavior of solids.

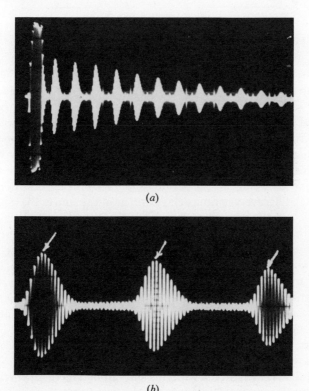

Fig. 3.10 (*a*) Applied pulse and echo train. (*b*) Echoes (expanded) showing individual cycles used in the time-delay measurement.

A significant improvement in the sensitivity of the pulse-echo method, especially for measuring changes in the delay time, was made by Cedrone and Curran.[34] Their procedure, known also as the "sing-around method" (and illustrated by the simplified diagram shown in Fig. 3.11), uses two transducers to serve as transmitter and receiver as in the pulse-transmission method.

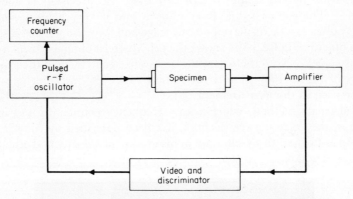

Fig. 3.11 Simplified block diagram of the sing-around method.

The essential idea here is to employ the signal from the receiving transducer in a feedback loop to trigger the pulse generator and establish the pulse-repetition rate. When the steady state condition is achieved, the period of the pulse-repetition frequency will be almost equal to the delay time of the specimen. This is a very precise method and with improvements has been made sensitive to *changes* in delay time of at least 0.1 ns. Counting the repetition frequency of the pulsed oscillator establishes a knowledge of the period of average time delay with remarkable precision. This average time delay consists of the sum of the true specimen delay, delay of the transducers and bonding material, and the electronic delays in the equipment, so there is the usual limitation on accuracy. The effect of the transducer-bond delay can be minimized with the use of two different-sized specimens, as described previously, and by the use of matched transducers and equal bond thicknesses on both specimens. There remains, however, the trigger-point delay (see Fig. 3.12), which occurs because the oscillator is not triggered at the instant of arrival of the pulse but at some fixed voltage level along the pulse. The difference is a source of timing error. With the technique used by Cedrone and Curran, an accuracy of 0.1 percent was achieved.

A difficulty is encountered with this sing-around method when specimens of very low loss are used. When the specimen attenuation is very small, the amplitudes of the directly transmitted pulse and the echo which results

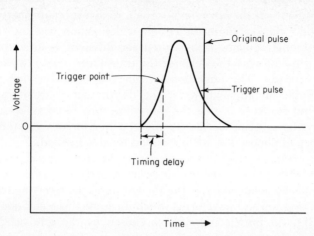

Fig. 3.12 The trigger-point delay is the time between the arrival of a pulse and the triggering of the next pulse. The timing delay results because the two do not exactly coincide.

from the first multiple reflection ($1\frac{1}{2}$ round trips) may be almost identical, and the discriminator circuit fails to differentiate between them. To overcome this deficiency, Meyers et al.[35] modified the sing-around method of Cedrone and Curran by the introduction of a gating circuit in the feedback path, as indicated in Fig. 3.13. The gate is closed at the time the oscillator pulses and remains closed for a preselected time, opening only to pass an echo which has been between the ends of the specimen. If the signal is the nth echo, then it will have traversed the specimen length $(2n - 1)$ times and will be attenuated accordingly. In this way the effective length of the specimen is increased, and short lengths of low-loss material

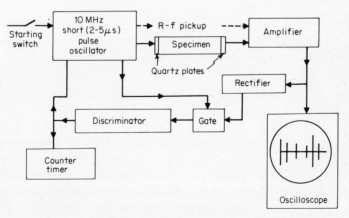

Fig. 3.13 Block diagram of the gated sing-around circuit.

may be studied. If the nth echo is chosen so that it is sufficiently reduced in amplitude, the discriminator circuit will function without ambiguity. The repetition frequency is measured for different echoes to eliminate electronic time delays—the delay of the transducer bond is accounted for as already noted. The limitation on selecting echoes is that for low n there be sufficient attenuation for the discriminator circuit to function properly and for large n, that the pulse rise time be of sufficiently short duration so retriggering occurs on the first quarter-wave of the rf cycle. An accuracy of almost 1×10^4 is claimed for this method.

A further improvement in the sing-around technique was developed by Forgacs.[36] He added a circuit to select a particular rf cycle within a preselected echo obtained by gating the output of the receiving transducer. The block diagram and waveforms which illustrate this technique are given in Fig. 3.14. A preselected echo is chosen by the echo-select circuit,

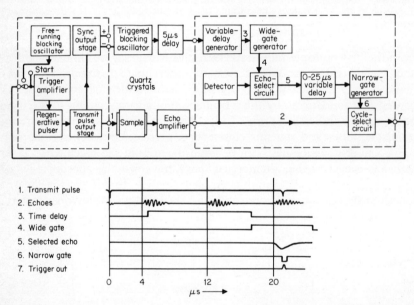

Fig. 3.14 Block diagram of the improved sing-around system including illustrative waveforms.

which is operated by the wide-gate generator. The sync-output stage triggers the blocking oscillator simultaneously with the transmitter signal to the sample. The blocking-oscillator output is delayed sufficiently in order to allow selection of the desired echo by gating the echo-select circuit. This signal is further delayed to control the narrow-gate generator in order to choose a particular cycle in the echo arriving at the cycle-select circuit. The output of this last circuit provides the triggering voltage to repeat the process. Monitoring the repetition frequency of the regenerative pulses, the duration of the time delay may be determined with a very high degree of precision. For observing a *change* in delay time, a sensitivity of 1 part in 10^7 has been achieved (for a 10-μs delay time, this corresponds to a detection of 0.01 ns change). The cycle that is selected and tracked should be located at the maximum amplitude of the echo in order to yield a maximum rise time and a minimum of jitter and drift. System stability, checked with a sample of single-crystal silver, was 5 parts in 10^8 for measurements made every 7 s over 4 min, 1 part in 10^7 for over 10 min.

For ceramic materials, the manner in which the acoustic pulse may be excited into the specimen is fairly limited to the use of piezoelectric devices, which require the transducer to be bonded to the material under study. This restricts the method to lower temperatures, since exposure of the transducer to high temperatures is limited by its Curie point, and transmission of the energy is further restricted by the response of the bond at elevated temperatures. For specimens of low acoustic loss, Krause[37] has devised a very useful approach to the problem. The basic arrangement is shown in Fig. 3.15a. A long sample is used so that the end to which the transducer is fastened may be kept out of the deleterious environment. The opposite end of the sample is notched over a short length, in order that it be entirely under uniform conditions. The time between the two returning echoes is quite small, and the method depends on a sufficiently precise measure of the small time difference. This effect is achieved (see Fig. 3.15b and c) by generating a reference signal delayed slightly (5 μs) behind the applied pulse and displayed on one of the channels of the dual-trace oscilloscope. The applied pulse and the two echoes are displayed on the other channel. The pulse-repetition rate is first adjusted so it is aligned with echo A, and then with echo B. The repetition frequency F is measured, in each instance, with the frequency counter. The differential delay is given simply by

$$\Delta T = \left(\frac{1}{F_A} - \frac{1}{F_B}\right) \tag{3.10}$$

as the fixed 5-μs delay subtracts out. The delay measurement of each individual echo is reported to have an accuracy of about 0.03 percent for a 30-μs delay. The difference in the delays between reflections A and B is smaller, about 0.1 percent for a 3-μs delay differential. A block diagram of this arrangement is shown in Fig. 3.16.

In an attempt to improve the direct pulse-echo technique for the purpose of determining *changes* in the propagation velocities (and hence elastic

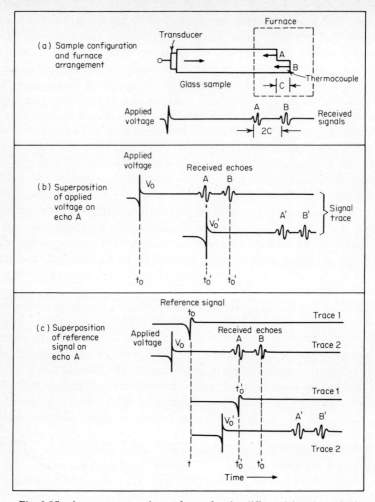

Fig. 3.15 Arrangement and waveforms for the differential path method.

constants), Chick[38] and his colleagues developed circuitry which permits detection of *changes* in travel times as small as 0.01 ns (10 ps). The approach is principally one of brute force, in which the pulse-echo techniques are pushed to their limits. They have been able to detect the very small *changes* in velocity which result from the application of a magnetic field to aluminium single crystals, and also to study the nonlinear properties of solids.[39] The method employed is shown in the block diagram of Fig. 3.17.

The 1-MHz crystal-controlled blocking oscillator provides the basic reference time base for the digital delay and also a 1-kHz repetition rate

The Determination of Velocity of Propagation 57

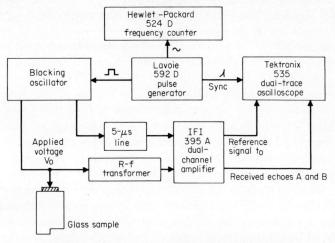

Fig. 3.16 Diagram of the differential path method.

trigger for the rf pulsed oscillator, which is obtained from the divider circuit. An energy pulse (3 to 10 MHz) is applied to the specimen, and the echo train (arising from the reflections at the specimen end) is detected in the time interval during which the rf pulsed oscillator is blocked by the broadband receiver. A signal from the digital delay permits selection of a particular echo in the received train of echoes, and the rf cycle gate permits selection of a particular cycle within the selected echo. The variation in time delay of the sonic energy within the specimen, as a function of some

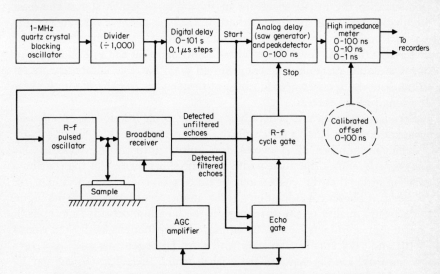

Fig. 3.17 Simplified block diagram of method employed by Chick et al.

externally applied parameter, is detected by the analog delay circuit, and the result read out on the high impedance meter, the output of which may also drive recorders. The sensitivity of the method is gained through a lever principle. If the *change* in time delay is, say, 1 ps/round trip of the energy pulse, then it is 10 ps/10 round trips and 100 ps/100 round trips. By selecting the nth echo (round trip), the *change* in the delay time is increased $n$fold. Hence a very small *change* in time delay may become detectable. This is at once the source of disadvantage of this method, for in requiring a large number of echoes, it places rather severe requirements upon the acoustic quality of the specimen. Advantages of this particular arrangement for measuring *change* in the time delay of a specimen are that the final readout is a dc voltage, which may be used to operate a recorder; there is a very short time constant in the readout, so that response time is at a minimum; the advantage of obtaining measurements of loss is preserved by viewing the decay of the entire echo train; and the method is readily made completely automatic.

While both greater accuracy and precision are achieved by employing the pulsed rf oscillator in the pulse-echo technique, in contrast to the dc pulse employed in the direct "time-of-flight" or pulse-transmission methods, the gain accrued is not without a compensating loss. This is of such a nature that it requires specimens of higher acoustic quality. The pulse-transmission methods have been routinely used with materials having a Q of a few hundred and even less than one hundred in some instances. The uncertainty in the data may be large, but some value may be assigned to the elastic property. In this latter method, the specimen responds to a frequency in the broad spectrum of the pulse to which it has the least attenuation. In the former method, it may be necessary to survey the carrier frequencies in order to find one which will be transmitted by the material if the Q is low. This adds the burden that the pulsed oscillator be of a stable variable-frequency design and that the laboratory invest in transducers to cover a wide range of frequencies to generate both shear and longitudinal waves.

Because attenuation is generally less at the lower frequencies, oscillators resonant down to 1 MHz are employed. However, for the echo to contain a sufficient number of rf cycles, the pulse width must necessarily be lengthened. Since the pulse width must be short relative to the specimen delay in order that the echoes may be time-resolved, longer specimens are required. These considerations force more severe requirements upon the acoustic quality of the specimens if the far more sensitive methods of acoustic interferometry are to be employed. Specimens with Q in excess of 10^3 and very small grain size, which are homogeneous and have very low porosity, are the types of material to which these techniques may be applicable.

3.4 Acoustic Interferometry

The variations of the pulse-echo techniques discussed so far are capable of excellent sensitivity, but of a more uncertain accuracy. They also lack the elegance of the methods to be described in this section. As with light, techniques based upon the behavior of wave interference were devised and, as was natural for such methods, they have the potential not only for attaining high sensitivity but for achieving a high degree of accuracy as well. For this reason, it becomes desirable to be able to evaluate the effect of the transducer and of the coupling material upon the measured time delay ("seal corrections"). When this effect is accounted for, an accuracy of 0.01 percent may be obtained. Two interferometric methods will be discussed. These are the phase-comparison and pulse-superposition methods. In the former method, the carrier frequency (which may be as low as 10 MHz or as high as 500 MHz) is used to establish the interference condition, and a stable rf oscillator is required for precise measurement of the specimen time delay. In the latter method, a pulse-repetition frequency establishes the interference condition, and the requirement is for stability of the pulse-repetition oscillator rather than for the rf oscillator. In both instances, seal corrections can be computed; however, the pulse-superposition technique is more suitable for determining the variation of time delay with, say, pressure since the seal correction can be made constant as the environment is changed. On the other hand, the phase-comparison method, when adapted for use with a buffer rod, has proven quite successful in measuring the velocity of very small specimens (1 to 2 mm long) using frequencies up to 500 MHz.[40]

3.4.1 Phase-Comparison Methods.
Among the earlier techniques for phase comparison is the double-pulse or double-gate method.[41] The basic circuit is shown in Fig. 3.18. Gates 1 and 2 are first opened simultaneously, and the output is viewed on the oscilloscope. The phase and amplitudes are adjusted for complete cancellation. For the measurement, gate 1 is opened and a signal is applied to the transducer, which results in the first echo train. The time at which gate 2 opens is delayed by one round-trip delay interval, so that the echoes are aligned as in Fig. $3.18 B_1$ and B_2. The carrier frequency is varied until a phase cancellation occurs, as shown in Fig. $3.18 B_3$, for all but the first echo of the first echo train. In fact, transients at the beginning and end of each echo will remain because of differences in pulse length and the rise and fall times of the two initiating pulses. Because the phase of the carrier within the pulse is a function of the carrier frequency, a sequence of nulls will be produced by varying the carrier frequency, causing it to sweep through conditions of destructive interference. This can be shown from the analysis, following Williams and Lamb.[41]

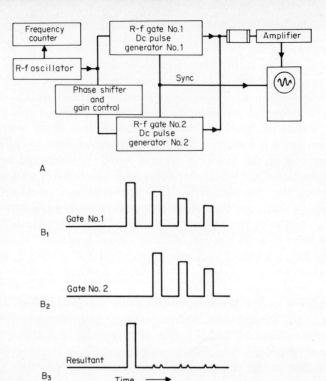

Fig. 3.18 Double-pulse, double-gate method: (A) Block diagram. (B_1) Pulses from gate 1. (B_2) Pulses from gate 2. (B_3) Resultant of phase adding these pulses for the out of phase condition.

Consider the received signal which arises from the second pulse to be of the form

$$A \sin \omega t \quad (3.11a)$$

where $A =$ amplitude constant
$\omega = 2\pi f$, where f is the carrier frequency

If the delay between the echoes is 2τ (the actual round-trip delay of the specimen and a phase shift due to the seal effect, Φ), then the received signal due to the first pulse is

$$A[\sin \omega(t - 2\tau) + 2\Phi] \quad (3.11b)$$

The output is the sum of (3.11a) and (3.11b)

$$A\{(\sin \omega t) + \sin[\omega(t - 2\tau) + 2\Phi]\}$$

or
$$2A[\sin(\omega t - \tau + \Phi)]\cos(\omega \tau - \Phi) \quad (3.11c)$$

From Eq. (3.11c), we note that the amplitude of the combined output can be made zero independently of t, by varying the carrier frequency ω. Then
$$\cos(\omega\tau - \Phi) = 0$$
from which
$$\omega\tau - \Phi = \frac{(2n+1)\pi}{2} \tag{3.12}$$
where n is a positive integer. Solving for τ, and expressing ω_n as $2\pi f_n$, the velocity is given by
$$v = \frac{2\pi f_n l}{n + \kappa_n} \tag{3.13}$$
where l is the length of the specimen, and κ, which includes the correction arising from phase shifts introduced at the transducer seal boundary, is given by
$$\kappa = 0.5 + \frac{\Phi_n}{\pi} \tag{3.14}$$

To determine v from a measured f, n must be known. This integer may be evaluated by determining a sequence of frequencies at which signal cancellation occurs. Then for any pair of consecutive frequencies, say f_n and $f_{(n+1)}$, we have
$$n = \frac{f_n}{\Delta f} - \frac{\kappa_n f_{(n+1)} - \kappa_{(n+1)} f_n}{\Delta f}$$

Since the phase shift $\Phi_n \neq \Phi_{(n+1)}$, the second term on the right may not be arbitrarily simplified. Substituting for κ, we can express the relation for n as a function of the phase correction,
$$n = \frac{f_n}{\Delta f} - 0.5 - \frac{f_{(n+1)}\Phi_n - f_n \Phi_{(n+1)}}{\pi \Delta f} \tag{3.15}$$

The integer n may be evaluated by dividing the nth frequency interval between the nth and $(n+1)$th frequencies at which cancellation occurs. By finding a series of frequencies at which the signal cancellation occurs, the measurement is cross-checked. Ambiguities in n may arise if the seal effect is not taken into account. An approximate expression of the phase shift for the arrangement described, using as thin a seal as possible (seal thickness approaches zero), is
$$\Phi_n = \pi \left[1 - \frac{2Z_t(f_n - f_0)}{Z_s f_0} \right] \tag{3.16}$$
where $f_0 =$ frequency at which the transducer is resonant
Z_t, $Z_s =$ acoustic impedances of the transducer and specimen, respectively

Using this relation for Φ_n and $\Phi_{(n+1)}$, Eq. (3.15) becomes

$$n = \frac{f_n}{\Delta f} - 1.5 - \frac{2Z_t}{Z_s} \qquad (3.17)$$

Because only two significant figures are generally needed for the correction (Z_t/Z_s), only a rough knowledge of the specimen velocity is required. An adequate determination may be made simply by measuring the distance between echoes, as they appear on the oscilloscope, and using the calibrated time base to determine the approximate round-trip time delay.

As an example[41] to illustrate the magnitude effect of the correction, consider a specimen of fused quartz with a length 1.8832 cm. The piezoelectric element is an X-cut quartz wafer (compressional mode) resonant at 10.082 MHz. The acoustic impedances are $Z_t = 15.15$ and $Z_s = 13.05$. An interval between two adjacent frequencies at which cancellation occurs is $\Delta f = 0.15320$ MHz and $f_n = 10.1928$ MHz. For these conditions, n is found to be 63. The correction due to the term containing Z_t/Z_s upon the velocity is found to be [Eq.(3.16)] 0.0255, so that ignoring the phase shift which occurs at the reflection would introduce an error of less than 0.05 percent in calculating the velocity. However, this correction cannot always be ignored in the determination of n without introducing an error of several percent. If the specimen has a low acoustic impedance, then the error in velocity which may arise by omission of this correction may approach 0.1 percent. Fortunately the acoustic impedance of most ceramic materials is large, so that ignoring the correction causes no significant error.

An interesting variation of this technique was used by McSkimin[42] in an arrangement for measuring the *change* in time delay requiring a very sensitive approach, and this is illustrated in Fig. 3.19. The advantages gained over the former method are (1) a simpler operating procedure is involved,

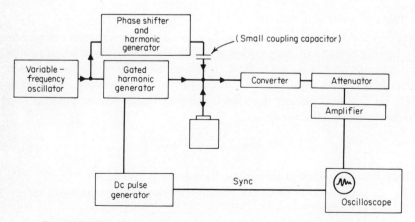

Fig. 3.19 Phase-comparison measuring circuit using carrier injection.

since only one gating circuit is employed, and (2) by using only one transducer (single-ended operation), a simpler and smaller specimen holder is used. This is of particular advantage when the specimen is to be placed in a pressure vessel or cryostat, where the available space is usually limited. It also simplifies the problem of devising electrical feed-throughs into the environmental chamber and eliminates other electrical difficulties. A disadvantage of the method is that the acoustic quality of the specimen must be considerably better than that required by the former method. In the former case a specimen which results in an echo train consisting of only two or three echoes is suitable for measurement.

The operation is similar to the former method in that a series of cancellations occur in the echo pattern when the carrier wave within an echo is out of phase with the carrier that is "leaked" in directly to the specimen continuously through the phase shifter–harmonic generator circuit. Use of the harmonic generator gives the added advantage that the phase shifter can function through 360°. The advantage of using the gated harmonic generator (gated at about 1 to 2 kHz), rather than a gated amplifier, is twofold. First, it allows the controlling oscillator to be of a low-frequency design and therefore more stable than if a high-frequency VFO (variable-frequency oscillator) were employed, and it further allows operation over a wide (20 to 180 MHz) frequency range. Second, it blocks the output of the oscillator during the gated period so that spurious signals are not introduced. This factor is quite crucial in the phase-comparison methods to be discussed later.

To follow the change in frequency, the dc pulse generator is adjusted so that the pulse width of the harmonic generator is kept sufficiently small in order to allow resolution of the echo train. The level of the amplitude wave leaked directly to the specimen is adjusted with the coupling capacitor. Variation of the rf oscillator frequency causes the echo pattern to go through a series of minima. At one of these minima, the phase of the leaked carrier wave is adjusted so that the echo pattern yields a sharp minimum. As the delay time of the specimen changes in response to the changes in its environment, the amplitude of the echo pattern builds up. The oscillator frequency is adjusted to minimize the amplitude of the echo pattern, and this frequency is measured at each new condition.

The change in time delay arises from two effects. First, the length changes as the specimen expands or contracts (for temperature variation) or shortens (for a pressure increase), and second, it changes in response to the change in the appropriate elastic modulus. The velocity may be simply expressed as

$$v = v_0 \left(\frac{f}{f_0}\right)\left(\frac{l}{l_0}\right) \tag{3.18}$$

where $f/f_0 =$ frequency ratio
$l/l_0 =$ length ratio

The subscript refers to the value at the initial (ambient) conditions of the experiment.

The ratio f/f_0 is determined directly from the experiment. The ratio l/l_0 must be obtained from a knowledge of the thermal expansivity. In the case of a variation in v due to a pressure change, the method of obtaining l/l_0 involves an integration employing the measured frequency-pressure data. These complications will be discussed in detail in a later section.

A very simple change in the arrangement alters the technique to what is sometimes referred to as the long-pulse method. One simply removes the path through which the radio frequency is leaked directly to the specimen, and varies the length of the applied pulse (and hence the width of the echoes). The sequence is shown in Fig. 3.20. In Fig. 3.20a, the echoes are resolved by employing a narrow pulse from the dc pulse generator. The pulse width is increased so that the returning echoes overlap, and the carrier wave within the echoes can then be made to interfere by varying the oscillator frequency. These effects are shown in Fig. 3.20b and c. The same effect as that achieved by the technique shown in Fig. 3.19 is accomplished with a simpler arrangement of equipment. The interfering pulse is introduced in this case by the simple expedient of widening the pulse sufficiently to overlap the echoes. A sensitivity to changes of 1 part in 10^5 has been achieved and an equal precision has been attained. For a specimen having a 1-μs time delay, this corresponds to detecting a change in time delay of 0.01 ns.

For very accurate measurements of absolute velocity, a buffer-rod technique employing phase comparison has been devised[43] which has proven very valuable, particularly in performing measurements upon very small specimens. Vitreous silica and quartz buffer rods have been used in the shape shown in Fig. 3.21A. The ends of the rod are polished flat to at least one wavelength of sodium light and parallel to 1 part in 10^4. The ends of the specimen should also be prepared to the same specifications if the full accuracy of the method is to be realized. The electronic circuitry is the same as shown in Fig. 3.19.

The transducer is bonded to the flat end of the buffer rod. If several rods are available, then a permanent bond employing a silver paste may be used. The specimen is attached to the beveled end with a suitable cement such as Dow Corning Resin 276-V9 (poly-α-methyl styrene). This material is sufficiently viscous at room temperature to transmit shear as well as longitudinal waves. The substance is worked, while warm, to produce a thin seal, and the assembly allowed to cool before taking measurements.

The appearance of the signal as viewed on the oscilloscope is shown in Fig. 3.21B_1 to B_3. Figure 3.21B_1 shows three sets of echo trains which arise from the applied pulse. The small arrows refer to the reflections which

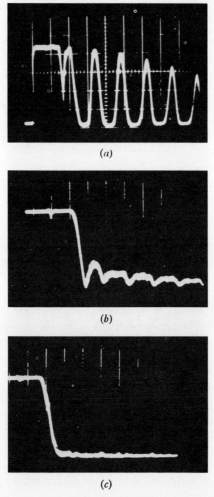

Fig. 3.20 (*a*) Applied pulse and return echoes; video output only is viewed. (*b*) Echoes overlapped; carrier-wave frequency slightly shifted from resonance. (*c*) Resonance condition; the carrier in the overlapped echoes are destructively interfering.

occur at the beveled end of the buffer rod, and are separated by the time delay in the rod. A part of the energy reaching the end of the rod is transmitted through the bond and into the specimen, where it is internally reflected. A portion of this reflected energy, following each round trip, is transmitted back into the buffer rod and is detected by the transducer to produce the series of specimen echoes in between the buffer rod echoes. The pulse width is lengthened to roughly half the time delay between the rod echoes, so that the strong interface reflections and the energy transmitted

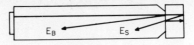

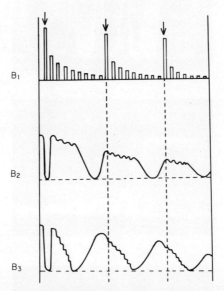

Fig. 3.21 (A) Reflections in buffer rod and sample and (B_1 to B_3) oscilloscope patterns viewed.

back from the specimen can phase-interfere. Then, by varying the oscillator frequency, a set of frequencies at which the energy in the specimen is *in phase*, but *out of phase* with the interface reflection, may be found. At these discrete frequencies, the signal goes through a sharp minimum, as shown in Fig. 3.21B_3. The propagation velocity of sound in the specimen is given by

$$v = \frac{2lf_n}{n + \dfrac{\Phi_n}{2\pi}} \quad (3.19)$$

where the notation is the same used previously. The integer n is determined by

$$n = \frac{f_n}{\Delta f} - \frac{\Phi_n f_{(n+1)} - \Phi_{(n+1)} f_n}{2\pi \Delta f} \quad (3.20)$$

The integer n can be determined, without ambiguity, using only the first term on the right if precautions are taken to keep the Φ terms small.

The phase shift Φ can be evaluated without any direct knowledge of the seal thickness. The computation involves the measured ratio of amplitudes of the signal applied to the specimen (noted by the arrows in Fig. 3.21B_1), and that of the first specimen echo, and the acoustic impedances of the specimen, buffer rod, and coupling material. A computer program for calculating the phase shift for different values of the amplitude ratio is given in Table 3.2. It is the work of Paul Glynn of the Bell Telephone Laboratories.

TABLE 3.2 Program for Computing Phase Angle Phi

```
2     READ,ZR,ZS,ZØ
3     DIMENSIØNBØ(1000),E(1000),PHI(1000),RATDB(1000),AØ(1000)
4     A=(ZR/ZS)−(ZS/ZR)
5     B=(A+(ZØ**2)/(ZR*ZS))−(ZR*ZS/ZØ**2)
6     C=(ZØ/ZS−ZS/ZØ)
8     DØ121=1,901
9     AØ(I)=(901−I)
92    BØ(I)=AØ(I)/10
10    E(I)=4./SQRTF(((A−B*(SINDF(BØ(I)))**2)**2)+(C*SINDF(2.*BØ(I)))**2)
11    PHI(I)=ATANDF(C*SINDF(2.*BØ(I))/(A−B*(SINDF(BØ(I)))**2))
12    RATDB(I)=20.*LØG10F(E(I))
13    DØ24N=1,801,200
14    PRINT15,ZR,ZS,ZØ
15    FORMAT(1H1,3X,"ZR=",F6.2,3X,"ZS=",F6.2,3X,"ZØ=",F6.2/" "4(2X,
    1 "RATIØ",3X,"PHI",5X,"BL",3X,"RATIØ",3X)/" "4(3X,"IN",21X,
    2 "ØF",4X)/" "4(3X,"DB",21X,"E",5X))
16    DØ24NP=1,46,5
17    DØ24M=1,5
18    I=M+NP+N−2
20    PRINT21,RATDB(I),PHI(I),BØ(I),E(I),RATDB(I+50),PHI(I+50),BØ
    1 (I+50),E(I+50),RATDB(I+100),PHI(I+100),BØ(I+100),E(I+100),
    2 RATDB(I+150),PHI(I+150),BØ(I+150),E(I+150)
21    FØRMAT(" "4(F8.3,2X,F5.2,3X,F5.2,1X,F6.4,2X))
24    CONTINUE

      STØP

      END
```

The necessary input data are the impedances of the buffer rod, specimen, and seal (ZS, ZR, ZØ, respectively). The program computes amplitude ratios of the applied signal and first echo [denoted by E(I)] for a series of seal thickness (BØ) and uses these to compute the phase shift (denoted by PHI). These are printed out along with the value of E(I) expressed in decibels (RATDB), which is the unit used in the measurement of the signal amplitudes. The value of PHI appropriate to the measured amplitude ratio is determined from this table and used in Eq. (3.19) for calculating the velocity. Impedance values used are 13.1 for a vitreous silica buffer rod, 15.1 for quartz compressional and 10.4 for quartz shear buffer rods, and 2.25 and 0.90 for compressional and shear in the bond material (Dow Corning Resin 276-V9).

3.4.2 Pulse Superposition. Pulse superposition differs from phase comparison in that the frequency of the applied pulses is crucial to the measurement, rather than the frequency, of the carrier. This technique is also due to McSkimin,[44,45] and has the operating advantage of greater energy in the return echoes. Figure 3.22 is the block diagram of the

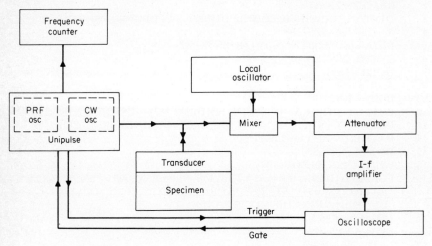

Fig. 3.22 Ultrasonic interferometer for pulse superposition.

electronic arrangement for pluse superposition (PSP). The critical component is the pulsed oscillator. The CW oscillator is of variable frequency design, with a range of 8 to 130 MHz, permitting the use of a wide range of transducers. The CW oscillator is pulsed by the pulse-repetition oscillator at a pulse-repetition frequency (PRF) of 1 MHz or less. The output of the pulsed oscillator consists of a sequence of rf pulses which drive the X- or Y-cut transducers. The returning specimen echoes are received by the same transducer. The signal is fed to a mixer stage, a calibrated attenuator, and a high-gain i-f amplifier-detector. The detected signal is applied to a Tektronix 535A oscilloscope, where the envelopes of the pulses are displayed. The PRF is accurately determined to six significant figures with the frequency counter. This method requires a very stable PRF, at least 1 part in 10^7.

In principal, the PRF is adjusted so that its period is equal to some integral submultiple of the delay time in the specimen. That is, the time delay between applied pulses (period of PRF) is exactly equal to an integral number of round trips in the specimen. When this condition is achieved, the applied pulses are *superimposed* upon the specimen echoes. If the integer is 1, every specimen echo will have an applied pulse superimposed upon it. If the integer is 2, the applied pulses will be superimposed upon every other echo, and so on. It is preferable to operate with this integer equal to unity, for then the greatest amount of energy is being impressed upon the specimen. At this condition, only the applied pulses are visible on the oscilloscope display. To observe the specimen echoes, a "window" is produced in the sequence of applied pulses by gating the pulsed oscillator. This is accomplished by applying a gating voltage from the oscilloscope to the pulsed oscillator. One now views the echoes and critically adjusts the PRF to maximize their amplitudes. When critically adjusted, the applied pulse is superimposed upon the echo, and the continuous wave in the pulse is phase-adding with the continuous wave in the echo. The relation between the measured time delay (reciprocal of the PRF) and the actual time delay in the specimen is given by

$$\tau = p\delta - \frac{p\xi}{2\pi f} + \frac{n}{f} \qquad (3.21)$$

In Eq. (3.21) τ is the measured period of the PRF at the interference condition, δ is the true time delay in the specimen, p is the integer discussed above, ξ is the phase shift introduced by the seal (transducer to specimen bond), f is the frequency of the CW oscillator, and n is an integer associated with the phasing between the continuous wave within the applied pulse and within the return echo. To understand the meaning of n, consider Fig. 3.23.

A constructive interference will occur every time a CW cycle in the pulse is exactly in phase with a CW cycle in the echo. A series of maxima may therefore be observed as the PRF is varied. These maxima will be separated by differences in the PRF corresponding to the period of the CW frequency. The maxima will be observed for each integral n for $n \leq 0 \leq n$. It is possible to determine the PRF for each $n = 0$, as described by McSkimin[45] or by an alternate procedure discussed below. Choosing the conditions $p = 1$ and $n = 0$, the true time delay in the specimen (time per round trip) is given by

$$\delta = \tau + \frac{\xi}{2\pi f} \qquad (3.22)$$

Generally, ξ is less than 1° for a properly prepared seal, and the continuous wave is of the order of 10^7 Hz. The correction is of the order of 10^{-10} or less, so that even for precise measurements this correction may fre-

70 Elastic Constants and Their Measurement

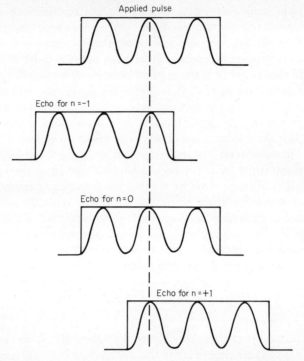

Fig. 3.23 Illustrating the meaning of the integer n in Eq. (3.21).

quently be ignored. The simple manner of dealing with these sources of systematic error is a major advantage of pulse superposition.

A further advantage arises from the fact that for CW frequencies near the transducer resonance frequency, the value of ξ is insensitive to changes of temperature and pressure.

The velocity of sound in the specimen is readily obtained once the specimen length l is known. The velocity is given by

$$v = \frac{2l}{\delta} \qquad (3.23)$$

Since δ is generally known to five or six significant figures, the greatest limitation on the accuracy of v is the measurement of the specimen length in the direction of sound propagation.

In order to employ the simple relation of Eq. (3.22), it is necessary to determine that the value of the integer n is zero without ambiguity. A useful method is to use two transducers resonant at, say, 20 and 30, or 20 and 40 MHz, and obtain a sequence of pulse-repetition frequencies with each transducer at which the signal maxima are observed. The true time delay of the specimen τ is independent of the transducer resonance frequency, and so the PRF will be the same for $n = 0$ in both cases (ignoring the effect of ξ, which is generally found to fall within the uncertainty of the PRF determination). From Eq. (3.21), we have for the two different transducers

$$\tau_1 = p\delta - \frac{p\xi_1}{2\pi f_1} + \frac{n_1}{f_1}$$

$$\tau_2 = p\delta - \frac{p\xi_2}{2\pi f_2} + \frac{n_2}{f_2}$$

and subtracting we have, for $p = 1$,

$$\tau_2 - \tau_1 = \frac{n_1}{f_1} - \frac{n_2}{f_2} - \frac{1}{2\pi}\left(\frac{\xi_1}{f_1} - \frac{\xi_2}{f_2}\right) \tag{3.24}$$

and from (3.24) it is clear that for $n = 0$, $\tau_2 = \tau_1$, ignoring the third term on the right, which is very small. The difference $(\tau_2 - \tau_1)$ will also equal zero whenever $f_2 n_1 = f_1 n_2$. This occurs for various n's removed from $n = 0$, and can be distinguished from the $n = 0$ condition by distortion in the shape of the echoes.

The interferometric methods described here have a unique advantage permitting both precise and accurate measurement of the delay times of acoustic waves in small specimens. However, there are certain limitations to the methods of which the experimenter should be cognizant. First we have the claim on specimen quality, which increases with decreasing specimen size. The reason for the latter lies in the nature of the method, which gains its sensitivity from the interference of a large number of carrier-wave cycles within each pulse. As specimen dimensions are diminished, the pulse widths must be made correspondingly smaller, and the number of CW cycles contained in a pulse is reduced in proportion. This may be overcome by raising the frequency of the carrier wave into the hundreds of megahertz region. This places a further stringent requirement on the specimen quality because it must be able to transmit these very high frequencies with only modest loss per round trip. This has been achieved with single crystals, using remarkably high frequencies. For polycrystalline materials, there is an added complication, and this is that the wavelength of the continuous wave be long in comparison with the grain size in order for the properties of the grains to be properly averaged. The minimum length of a polycrystalline material is therefore subjected to these two bounds. The smallest, practical specimens for polycrystalline ceramic is about 5 mm, and a specimen length of 10 to 15 mm is nearly ideal.

A second feature involves the rather stringent requirements on specimen preparation. For maximum sensitivity, the specimen's surfaces should be optically polished to a high degree of flatness and parallelism, an expensive and time-consuming procedure. This requirement is dictated by the following circumstances. For a specimen 1 cm long, with a velocity of propagation of, say, 10 km/s, the wavelength of a 30-MHz signal is about 33 μ. The method is sufficiently sensitive to detect phase differences of less than 1°, so that a path difference of 33/360, or about 0.3 μ in a 1-cm length, resulting from the two opposite surfaces of the specimen not being parallel, will result in noticeable phase shifts and an increased uncertainty in the delay-time measurement. Similarly, the surface roughness must also be made as small as possible, since (1) the specimen length must be known to five significant figures if the velocity is to be known to 1 part in 10^4, (2) scatter of the acoustic wave, and hence signal loss arising from a nonsmooth surface, must be eliminated, and (3) an optical polish is required in order to determine whether the surfaces are sufficiently parallel.

It was inevitable, with the continuing improvements in electronic designs, that systems become automated, and there are applications where such a system, capable of following rapid changes in propagated velocity (as through a phase transition), is highly desirable. One approach to automation is shown in Fig. 3.24 as applied to the pulse-superposition method.[46] The upper part of Fig. 3.24 is the basic pulse-superposition circuitry, with the exception that a frequency synthesizer is employed to provide the pulse-repetition frequency. The use of the frequency synthesizer is desirable in this application because of its singular capabilities

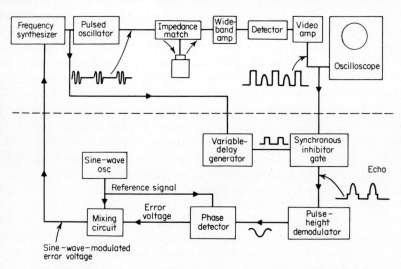

Fig. 3.24 Block diagram of the automatic pulse-superposition method.

with regard to stability and precision. The lower portion of Fig. 3.24 shows the arrangement of components added to achieve automatic control of the measurement. In this arrangement, the pulse-superposition method is operated in the $p = 2$ condition [refer to Eq. (3.21)]. The received pattern is applied to an inhibitor gating circuit, which suppresses the applied pulse, and the resultant echo pattern thus achieved, with a low-frequency sine wave superimposed on it, is amplitude-detected via the pulse-height demodulator. The output of the demodulator is applied to the quadrature phase detector which yields an output only when the phase of the signal is other than 90°. If the properties of the materials studied undergo a change, then the original setting no longer provides the maximum echo signal, because the echoes will either be arriving earlier or later with respect to the initial round-trip time through the specimen. These will now fall on the high- or low-frequency side of the initial pulse-repetition frequency. The effect is to shift the phase seen by the quadrature phase detector to either 0° or 180°, producing a positive or negative error signal. This error signal is mixed with a low frequency to produce the correcting voltage that is used to alter the output of the frequency synthesizer to the new value appropriate to the changed velocity. For complete automation, the output of the frequency synthesizer may also be fed through an analog-to-digital converter and then to a printer, allowing the operator to obtain a permanent record of the data completely automatically.

3.4.3 Transducers and Coupling Materials.

The most generally used piezoelectric materials for these methods are the various cuts of quartz transducers in the form of a wafer $\frac{1}{2}$ to 1 cm in diameter. For the phase-comparison methods employing harmonic generators, the transducers are polished for harmonic operation. A variety of orientations are available for generating shear and longitudinal waves, but the X-cut for longitudinal and Y- or AC-cut for shear types are often used. The AC-cut is sometimes preferred because it is less highly coupled and therefore generates a purer shear mode than does the Y. It is also thicker for a given frequency and less fragile than the Y-cut transducer. For frequencies above 40 MHz, quartz must be ground exceedingly thin, and so tourmaline is preferred, although it is more expensive.

To attach the transducer, a variety of materials have been employed, ranging from sealing wax to epoxy cements. For experiments at low temperatures or high pressures, the material should be capable of yielding to relieve stresses that may arise and crack the transducer. Should the transducer break, the experiment need not end so long as the pieces do not move relative to one another. At elevated temperatures, the seal represents the principal obstacle, since it is the failure of the seal that limits much of the high-temperature work. Fisher and Renken[47] have reported on a seal composed of a mixture of water glass and calcium carbonate useful to about 600°C. In general, the methods of interferometry have been applied to studies at low temperatures where the seal does not represent as critical a problem. Among several of the materials suitable to couple the transducer to the specimen are the following:

1. Ordinary waxes (paraffins), beeswax, sealing wax
2. Light machine oils (good for shear waves, too, at temperatures below about $-10°C$)
3. Salol (phenylsalicylate—excellent for shear waves up to 43°C)
4. Dow Corning Resin 276-V9 Blend 288 (poly-α-methyl styrene); useful for both longitudinal and shear waves from $+80°C$ to $-80°C$
5. Nonaq stopcock grease, which is suitable for both longitudinal and shear waves from 60°C to $-270°C$

The choice of a suitable bonding material remains largely a matter of trial and error. For example, a variety of soldering flux has been found suitable for coupling acoustic energy up to 250°C.

The quality of the seal is of utmost importance. A very thin seal, free of bubbles and prepared between clean surfaces, represents the optimal condition. This is achieved in several ways, depending upon the nature of the material used. For those substances which are thermal-setting, a simple jig to squeeze out any excess material and employing a load of about 25 psi on the transducer has produced uniformly good bonds. To prepare the salol seal, the material is melted and injected between transducer and specimen to form a thin film. A small seed crystal is placed at the edge of the transducer to initiate crystallization from a single point. With this seal, it is important that the salol crystal forming the bond have a single orientation. Salol forms a bond of excellent acoustic quality, but a successful seal is prepared with difficulty. Some of the materials require time for curing, and this is best done with a specimen-transducer assembly held firmly in the jig.

While the bonds described are fine for studies at or near room temperature, and are more than adequate for measurements performed at elevated pressure, they are unsatisfactory at elevated temperatures. To overcome this, a lapped bond, or—euphemistically—a no-bond bond has been devised. This requires very careful lapping of the specimen to the transducer (for pulse-superposition-type measurements) or to the buffer rod (for phase-comparison-type measurements). The surfaces to be wrung together must each be made optically flat so that they adhere to each other. For phase-comparison measurements, this may be facilitated by lapping the prepared surfaces of the buffer rod and specimen together, using an 0.05-μm grit, until they are equally flat and parallel and can be wrung together. To accomplish the same thing for pulse-superposition measurements is considerably more difficult because the transducer is so thin and fragile. In this instance, wringing together of the transducer and optically flat specimen surfaces may be readily accomplished by using a volatile liquid such as trichlorethylene as an aid. A drop of the liquid is placed on the specimen surface and the transducer put on the liquid. The transducer is pressed on the specimen and worked on. Wringing the two surfaces together is accomplished as the liquid evaporates.

3.4.4 Measurements at High Pressure and High Temperature. An important application of acoustic interferometry has been in the determination of the elastic behavior of solids under high pressure. The method is particularly suited for such studies because (1) its sensitivity permits a precise determination of the pressure derivatives of elastic constants and (2) the entire assembly of specimen and transducer is small enough to fit into the limited space of the pressure cell. The determination of the variation of sound velocity with pressure is an important experiment because it yields information concerning the manner in which the vibrational modes of a solid change with volume. Such data are extremely important in evaluating theories concerning the general equation of state of solids, and in elucidating theories concerning the nature of the forces bonding solids. A unique advantage results from the sensitivity of the method, which means that experiments need not be run at very high pressures in order to determine the values of the pressure derivatives of the elastic properties. Indeed, measurements to pressures of only 3 or 4 kbar are sufficient to determine pressure derivatives of velocity. This has the concurrent advantage of allowing the experiments to be run under truly hydrostatic pressure conditions, using less unwieldy and less expensive pressure-generating equipment.

As noted earlier, the primary data obtained from these time-delay measurements are the frequency ratios and the velocities calculated from Eq. (3.18). To repeat,

$$v = v_0 \left(\frac{f}{f_0}\right)\left(\frac{l}{l_0}\right)$$

It is necessary to know how l/l_0 is changing with pressure with a precision sufficient to define the change in v consistent with the precision in determining f/f_0. The difficulty is that the length ratio is determined by the compressibility, which is, itself, a function of pressure. Cook[48] has shown how this difficulty may be overcome, by actually using the experimental data. The details are as follows. The compressibility for an isotropic solid is given by

$$-\frac{1}{V}\frac{dV}{dP} = \chi_T \qquad (3.25)$$

where the subscript indicates the isothermal compressibility. In terms of length, Eq. (3.25) becomes

$$-\frac{3}{l}\frac{dl}{dP} = \chi_T$$

If we define a parameter $S = (l_0/l)$, the above becomes

$$\frac{3}{S}\frac{dS}{dP} = \chi_T \tag{3.26}$$

Now the compressibility calculated from velocity data is adiabatic, and to convert we use the relation

$$\chi_T = \chi_S(1 + T\alpha\gamma) = \chi_S(1 + \Delta)$$

where γ is the Grüneisen constant and can be computed from

$$\gamma = \frac{\alpha B_S}{C_P \rho} \tag{3.27}$$

where $\alpha =$ volume expansivity
$B_S =$ adiabatic bulk modulus
$C_P =$ specific heat at constant pressure

Further, the compressibility is given by $\{\rho[v_l^2 - (4/3)v_s^2]\}^{-1}$, so that Eq. (3.26) becomes

$$\frac{dS}{S} = \frac{(1+\Delta)dP}{\rho(3v_l^2 - 4v_s^2)} \tag{3.28}$$

The denominator can be written

$$\rho_0 S \left\{ 3\left[v_{0l}\left(\frac{f}{f_{0l}}\right)\right]^2 - 4\left[v_{0s}\left(\frac{f_s}{f_{0s}}\right)\right]^2 \right\}$$

and (3.28) becomes

$$\frac{l_0}{l} = 1 + \frac{(1+\Delta)}{\rho_0} \int_0^P \frac{dP}{3[v_{0l}(f_l/f_{0l})]^2 - 4[v_{0s}(f_s/f_{0s})]^2} \tag{3.29}$$

Integration of Eq. (3.29) is based upon the linear behavior of the frequency ratio-pressure data. In that case, the denominator can be expressed in the form,

$$(A + BP)^2$$

which, upon expansion, is approximated as $A + BP$. This approximation holds, since the B^2 term is about six orders of magnitude smaller than the P^2 term. In this case (3.29) integrates directly to

$$\frac{l_0}{l} = \frac{1}{B} \ln(A + BP)\big]_0^P \tag{3.30}$$

from which the length ratio may be computed. The density at any pressure may now also be computed.

The variation of the velocity with pressure can be calculated from Eq. (3.18), and the slopes then defined. The values of dv/dP may now be used to calculate the pressure derivatives of the elastic moduli. Of most interest are the pressure derivatives of the shear modulus, bulk modulus, and Poisson's ratio. The equations, obtained by direct differentiation of the modulus-velocity relationships, are

Shear modulus:
$$G = \rho v_s^2$$
$$\left(\frac{dG}{dP}\right)_T = 2\rho v_s \left(\frac{dv_s}{dP}\right) \tag{3.31}$$

Bulk modulus:
$$B_S = \rho(v_l^2 - \tfrac{4}{3}v_s^2)$$
$$\left(\frac{dB_S}{dP}\right)_T = 2\rho\left[v_l\left(\frac{dv_l}{dP}\right) - \frac{4}{3}v_s\left(\frac{dv_s}{dP}\right)\right] + (1 + T\alpha\gamma) \tag{3.32}$$

Poisson's ratio:
$$\sigma = \frac{1}{2}\left\{1 - \left[\left(\frac{v_l}{v_s}\right)^2 - 1\right]^{-1}\right\}$$
$$\left(\frac{d\sigma}{dP}\right)_T = 2(1 - 2\sigma)(1 - \sigma)\left(\frac{d\ln v_l}{dP} - \frac{d\ln v_s}{dP}\right) \tag{3.33}$$

Using these relationships, the pressure derivatives may be precisely defined. As an example, the data for polycrystalline alumina[2] are shown in Table 3.3.

An alternate method of calculating dB/dP has been derived by Thurston.[49] It differs from the calculation described above in that it yields the value dB/dP, while that of Cook yields $\Delta B/\Delta P$, or the slope of the chord (average slope over a pressure interval ΔP). For highly incompressible materials these are essentially the same, but for compressible materials, the experimental data may exhibit curvature, in which case the calculation described by Thurston should be used (see Chap. 7).

With the development of the lapped bond described earlier, it has become possible to perform the measurements described in this section under conditions of both elevated pressure and temperature. For temperatures in excess of the α to β quartz transition, the phase-comparison method, which employs a buffer rod, is advantageous, for then a sufficiently long rod can be used in order to keep the transducer at the low-temperature end of the furnace. This approach has been successfully applied by Spetzler[50] in measurements performed with the simultaneous application

TABLE 3.3 Longitudinal and Shear Velocity for Polycrystalline Alumina as Functions of Pressure at 25°C and at −78.5°C

Pressure, bars	$\dfrac{l_0}{l}$	Frequency ratio		Velocity, km/s	
		Longitudinal	Shear	Longitudinal	Shear
25°C					
1	1.0000000	1.000000	1.000000	10.845	6.3730
500	1.0000658	1.000305	1.000239	10.848	6.3741
1,000	1.0001325	1.000611	1.000479	10.850	6.3752
1,500	1.0001982	1.000917	1.000719	10.853	6.3763
2,000	1.0002648	1.001223	1.000958	10.855	6.3774
2,500	1.0003304	1.001528	1.001198	10.858	6.3785
3,000	1.0003969	1.001834	1.001438	10.861	6.3796
3,500	1.0004624	1.002140	1.001677	10.863	6.3807
4,000	1.0005300	1.002446	1.001917	10.866	6.3818
−78.5°C					
1	1.0000000	1.000000	1.000000	10.880	6.4025
500	1.0000656	1.000300	1.000230	10.883	6.4035
1,000	1.0001317	1.000600	1.000461	10.885	6.4046
1,500	1.0001973	1.000900	1.000691	10.888	6.4057
2,000	1.0002632	1.001200	1.000922	10.890	6.4067
2,500	1.0003287	1.001499	1.001152	10.893	6.4078
3 000	1.0003947	1.001799	1.001383	10.895	6.4088
3,500	1.0004601	1.002099	1.001613	10.898	6.4099
4,000	1.0005254	1.002399	1.001844	10.900	6.4109

of both pressure and temperature. The quartz transducer was cemented to one end of the rod with epoxy, and the specimen was wrung onto the other end. Using this approach, Spetzler and his colleagues have been able to obtain measurements defining the equations of state of several materials over a wide pressure-temperature regime, in contrast to prior measurements which could define the parameters for the isothermal equation of state only. A detail of their furnace and specimen assembly is shown in Fig. 3.25.

Methods for obtaining precise measurements of sound velocity continue to undergo change and development. As noted in the beginning of this chapter, not all the techniques in use have been described here, and new approaches are introduced as developments in electronic circuitry make these approaches feasible. The serious worker is advised to refer to journals such as the *Journal of Applied Physics*, the *Journal of the Acoustic Society of America*, and the *Review of Scientific Instruments* if a desire to keep up with the continuing progress in this area is to be satisfied.

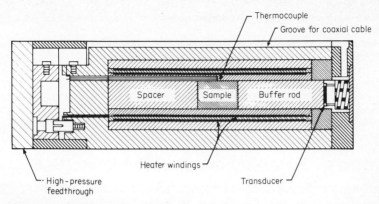

Fig. 3.25 Furnace design and specimen assembly for use in a pressure vessel to perform elastic property measurements at elevated temperature and pressure simultaneously.

REFERENCES

1. Schreiber, E.: Elastic Moduli of Single Crystal Spinel at 25°C and to 2 Kbar, *J. Appl. Phys.*, **38**(6):2508 (1967).
2. Schreiber, E., and O. L. Anderson: Pressure Derivatives of the Sound Velocities of Polycrystalline Alumina, *J. Am. Ceram. Soc.*, **49**(4):184 (1966).
3. Schreiber, E., and N. Soga: Elastic Constants of Hot-Pressed Silicon Carbide, *J. Am. Ceram. Soc.*, **49**(6):342 (1966).
4. McSkimin, H. J.: Ultrasonic Methods for Measurement, in W. P. Mason (ed.), "Physical Acoustics," vol. 1, pt. A, Academic, New York, 1964.
5. Anderson, O. L., and R. Liebermann: Sound Velocities in Rocks and Minerals, *Vesiac State of the Art Report #7885-4-x*, Willow Run Laboratories, University of Michigan, Ann Arbor, 1966.
6. Hughes, D. S., W. L. Pondrom, and R. L. Mims: Transmission of Elastic Pulses in Metal Rods, *Phys. Rev.*, **75**(10):1552 (1949).
7. Hughes, D. S., and H. J. Jones: Variation of the Elastic Moduli of Igneous Rocks with Pressure and Temperature, *Bull. Geol. Soc. Am.*, **61**:843 (1950).
8. Hughes, D. S., and J. H. Cross: Elastic Wave Velocities at High Pressures and Temperatures, *Geophys.*, **16**:577 (1951).
9. Hughes, D. S., and C. Maurette: Elastic Wave Velocities in Granites, *Geophys.*, **21**:277 (1956).
10. Hughes, D. S., and C. Maurette: Dynamic Elastic Moduli of Iron, Aluminum, and Fused Quartz, *J. Appl. Phys.*, **27**:1184 (1956).
11. Araki, J.: On the Propagation of Elastic Waves in Anisotropic Media I, *Mem. Ehime Univ.*, Sec. III (Eng.), **5**:83 (1965).
12. Araki, J.: On the Propagation of Elastic Waves in Anisotropic Media II, *Mem. Ehime Univ.*, Sec. III (Eng.), **5**:99 (1965).
13. Birch, F.: The Velocity of Compressional Waves in Rocks to 10 Kilobars, Part 1, *J. Geophys. Res.*, **65**:1083 (1960).
14. Mattaboni, P., and E. Schreiber: Method of Pulse Transmission Measurements for Determining Sound Velocities, *J. Geophys. Res.*, **72**:5160 (1967).
15. Mason, W. P.: "Physical Acoustics and the Properties of Solids," chap. 2, Van Nostrand, Princeton, N.J., 1958.

16. Bancroft, D.: The Velocity of Longitudinal Waves in Cylindrical Bars, *Phys. Rev.*, **59**:588 (1941).
17. Eros, S., and R. Reitz: Elastic Constants by the Ultrasonic Pulse Echo Method, *J. Appl. Phys.*, **29**:683 (1958).
18. Redwood, M.: Experiments with the Electrical Analog of a Piezoelectric Transducer, *J. Acoust. Soc. Am.*, **36**:1872 (1964).
19. Mason, W. P., and H. J. McSkimin: Attenuation and Scattering of High Frequency Sound Waves in Metals and Glasses, *J. Acoust. Soc. Am.*, **19**:464 (1947).
20. Pellam, J. R., and J. K. Galt: Ultrasonic Propagation in Liquids: I. Application of Pulse Technique to Velocity and Absorption Measurements at 15 Megacycles, *J. Chem. Phys.*, **14**:608 (1946).
21. Huntington, H. B.: Ultrasonic Measurements on Single Crystals, *Phys. Rev.*, **72**:321 (1947).
22. Lazarus, D.: The Variation of the Adiabatic Elastic Constants of KCl, NaCl, CuZn, Cu and Al with Pressure to 10,000 Bars, *Phys, Rev.*, **76**:545 (1949).
23. Galt, J. K.: The Mechanical Properties of NaCl, KBr, KCl, *Phys. Rev.*, **73**:1460 (1948).
24. Bacon, R., and C. S. Smith: Single Crystal Elastic Constants of Silver and Silver Alloys, *Acta Metal.*, **4**:337 (1956).
25. Overton, Jr., W. C.: Ultrasonic Measurements in Metallic Beryllium at Low Temperatures, *J. Chem. Phys.*, **18**:113 (1950).
26. Miller, R. O., and C. S. Smith, Pressure Derivatives of the Elastic Constants of LiF and NaF, *J. Phys. Chem. Sol.*, **25**:1279 (1964).
27. Bartels, R. A., and D. E. Schuele: Pressure Derivatives of the Elastic Constants of NaCl and KCl at 295°K and 195°K, *J. Phys. Chem. Sol.*, **26**:537 (1965).
28. Daniels, W. B., and C. S. Smith: Pressure Derivatives of the Elastic Constants of Copper, Silver, and Gold to 10,000 Bars, *Phys. Rev.*, **111**:713 (1958).
29. Bell, J. F. W.: The Velocity of Sound in Metals at High Temperatures, *Philos. Mag.*, **2**:1113 (1957).
30. Daniels, W. B., and C. S. Smith: The Pressure Variations of the Elastic Constants of Crystals, in "Physics and Chemistry of High Pressures," pp. 50–63, Gordon and Breach, New York, 1963.
31. McSkimin, H. J.: Ultrasonic Measurement Techniques Applicable to Small Solid Specimens, *J. Acoust. Soc. Am.*, **22**:413 (1950).
32. May, J. E.: Precise Measurement of Time Delay, *IRE Nat. Conv. Rec.*, **6**, pt. 2:134 (1958).
33. Papadakis, E. P.: Ultrasonic Attenuation and Velocity in Three Transformation Products in Steel, *J. Appl. Phys.*, **35**:1474 (1964).
34. Cedrone, N. P., and D. R. Curran: Electronic Pulse Method for Measuring the Velocity of Sound in Liquids and Solids, *J. Acoust. Soc. Am.*, **26**:963 (1954).
35. Myers, A., L. Mackinnon, and F. E. Hoare: Modifications to Standard Pulse Techniques for Ultrasonic Velocity Measurements, *J. Acoust. Soc. Am.*, **31**:161 (1959).
36. Forgacs, R. L.: Improvements in the Sing-Around Technique for Ultrasonic Velocity Measurements, *J. Acoust. Soc. Am.*, **32**:1697 (1960).
37. Krause, J. T.: Differential Path Method for Measuring Ultrasonic Velocities in Glasses at High Temperatures, *J. Acoust. Soc. Am.*, **35**:1 (1963).
38. Chick, B. B., G. P. Anderson, and Rohn Truell: Improvement in Ultrasonic Velocity Measurement Techniques, *Final Tech. Rep.* (*Contract NOw 66-0105d*), Oct., 1966, Metals Research Laboratory, Brown University, Providence, R.I.
39. Hikata, A., B. Chick, and C. Elbaum: Ultrasonic Studies of the Non-Linear Properties and of the Deformation of Solids, *Tech. Rep.*, *AF ML-TR-65-56*, Air Force Materials

Laboratory, Research and Technical Division, Air Force Systems Command, Wright-Patterson AFB, Ohio, 1965.
40. McSkimin, H. J.: Measurement of Ultrasonic Wave Velocities for Solids in the Frequency Range 100 to 500 Mc., *J. Acoust. Soc. Am.*, **34**:404 (1962).
41. Williams, J., and J. Lamb: On the Measurement of Ultrasonic Velocity in Solids, *J. Acoust. Soc. Am.*, **30**:308 (1958).
42. McSkimin, H. J.: Elastic Moduli of Single Crystal Germanium as a Function of Hydrostatic Pressure, *J. Acoust. Soc. Am.*, **30**: 314(1958).
43. McSkimin, H. J.: Measurement of Elastic Constants at Low Temperatures by Means of Ultrasonic Waves—Data for Silicon and Germanium Single Crystals, and for Fused Silica, *J. Appl. Phys.*, **24**:988 (1953).
44. McSkimin, H. J.: Pulse Superposition Method for Measuring Ultrasonic Wave Velocities in Solids, *J. Acoust. Soc. Am.*, **33**:12 (1961).
45. McSkimin, H. J.: Analysis of the Pulse Superposition Method for Measuring Ultrasonic Wave Velocities as a Function of Temperature and Pressure, *J. Acoust. Soc. Am.*, **34**:609 (1962).
46. Mattaboni, P.: Automatic Frequency Control System for the Measurement of Rapid Changes of Ultrasonic Wave Velocities in Solids, *Prog. Rep.* AF 44620-68-C0079, Air Force Cambridge Research Laboratories, Bedford, Mass., Jan. 1970.
47. Fisher, E. S., and C. J. Renken: Coupling Cements for Ultrasonic-Wave Velocity Measurements at High Temperatures, *J. Acoust. Soc. Am.*, **35**:1055 (1963).
48. Cook, R. K.: Variation of Elastic and Strains with Hydrostatic Pressure: A Method for Calculation from Ultrasonic Measurements, *J. Acoust. Soc. Am.*, **29**:445 (1965).
49. Thurston, R. N.: Effective Elastic Coefficients for Wave Propagation in Crystals Under Stress, *J. Acoust. Soc. Am.*, **37**:348 (1965).
50. Spetzler, H.: Equation of State of Polycrystalline and Single Crystal MgO to 8 Kilobars and 800°K, *J. Geophys. Res.*, **75**:2073 (1970).

CHAPTER FOUR

Dynamic Resonance Method for Measuring the Elastic Moduli of Solids

4.1 Introduction

Here we will be concerned with the problem of determining the elastic moduli of solids from their mechanical resonance frequencies. The type of vibration may be longitudinal (or extensional), flexural (or transverse), or torsional. The first two kinds supply information about Young's modulus and the last gives the shear modulus.

The exact solution of the three-dimensional form of the differential equations of motion is extremely difficult, and from the experimentalist's point of view, it is desirable to choose a geometry which provides boundary conditions that allow for a reasonably exact solution. For this reason, only the problems related to the vibration of specimens of rectangular or cylindrical shapes will be dealt with here. The analogous problem of dealing with objects of spherical shapes will be discussed in Chap. 5.

The dynamic resonance method, which originated from the early measurements by Ide[1] in 1935, is based on a standing-wave phenomenon. When the specimen is undergoing longitudinal or torsional vibration, the length of the specimen contains an integral number n of half-wavelengths, or

$$l = \frac{n\lambda}{2} \tag{4.1}$$

The velocity of the wave is then expressed by

$$v = \lambda f = \frac{2lf}{n} \tag{4.2}$$

where f is the resonant frequency of any mode of vibration. However, as described later, this simple formula does not apply to flexural resonance since the nodes are not at the quarter points, as simple standing waves would require.

The dynamic resonance method has a definite advantage over static methods of measuring elastic moduli because the low-level alternating stress does not initiate anelastic processes such as creep or elastic hysteresis. Under such a low stress, the assumptions of the mathematical theory are well fulfilled. The great improvement of the measuring system made at the National Bureau of Standards, especially by Spinner and Tefft,[2] made it possible to experimentally check the solutions for vibrating bodies based upon the theory of elasticity.

In the following sections, the equations relating the resonance frequency and elastic moduli will be described. Then the description of the measuring system and the method for identification of the modes of vibration, along with an example of calculating elastic moduli from the resonant frequencies, will follow. Since the dynamic resonance method is suitable for determination of the elastic moduli at high temperatures, some detailed description about such measurements will be given. In the last section, several modified resonance methods of measuring the elastic moduli of small specimens will be described.

4.2 Vibration of Cylindrical and Rectangular Bars

The theory of free vibrations of solid bodies requires the integration of the equations of vibratory motion in accordance with prescribed boundary conditions of stress or displacement. Theoretical expressions of longitudinal and torsional modes of vibrations for bars and cylinders were derived long ago, for example by Pochhammer[3] in 1876. However, exact solutions of the equations were not obtained until the 1940's, even for the simple cylindrical case. An exact solution for the relatively simple geometry of a bar with a rectangular cross section is still to be made; as a consequence, only approximate solutions exist. The improvements and refinements in measuring techniques have made it possible to verify the validity of these approximations, and as a result the dynamic resonance method has become a standard technique for determining the elastic constants of solids. In the following section, we deal with the problem concerning the accurate correlation of the measured resonance frequencies and the elastic constants.

4.3 Torsional Vibration

The general equation which relates the shear modulus G and the torsional resonant frequency f_G is

$$G = \frac{4\rho R l^2 f_G^2}{n} \tag{4.3}$$

where l = length of the specimen
ρ = density
R = a shape factor depending upon the shape of the cross section of the specimen
n = an integer which is unity for the fundamental mode, two for the first overtone, etc.

This equation applies to rods where the cross section is circular, square, or rectangular—the geometry being taken account of by the shape factor R.

Equation (4.3) and the equation describing the shape factor R can be derived from the differential equations for torsional vibration of the bar (as discussed later). The equation for the cylinder is both exact and simple to solve because the factor R which involves the polar moment of inertia of the cross section is unity and is independent of the diameter and length. On the other hand, the equation for the rectangular bar is very complicated and requires some approximations for its solution. Here, the solutions given by Davies[4] are quoted.

The differential equations of motion for *infinitely thin* bars of free ends are in the form

$$\frac{\partial^2 \Psi}{\partial t^2} = \frac{G}{\rho R} \frac{\partial^2 \Psi}{\partial z^2} = \frac{v_s^2}{R} \frac{\partial^2 \Psi}{\partial z^2} \tag{4.4}$$

where Ψ = angular displacement at time t at distance z from the origin
v_s = velocity of propagation of torsional waves in an infinitely thin cylindrical bar

As noted above, $R = 1$ for a cylinder. For an *elliptic* cross section with semiaxis $a/2$ and $b/2$, $1/R$ is given by $4a^2b^2/(a^2+b^2)^2$. If the cross section is *rectangular* with sides a and b ($a > b$, $s = a/b$), $1/R$ is given by (Davies[4])

$$\frac{1}{R} = \frac{3}{4(1+s^2)}\left[\frac{16}{3} - \frac{2}{b}\left(\frac{4}{\pi}\right)^5 \sum_{p=0}^{\infty} \frac{1}{(2p+1)^5} \tanh(2p+1)\frac{\pi s}{2}\right] \quad (4.5)$$

When $s = 1$, the cross section is a square, and $1/R$ becomes 0.8435.

When the length of the bar becomes short in comparison with its cross-sectional dimensions, Eq. (4.5) does not hold explicitly because it is based on the theory of vibration of an infinite bar. For a short bar, the following correction must be made (Davies[4]):

$$\frac{R'}{R} = 1 + \left(\frac{n\pi\mu}{l}\right)^2 \quad (4.6)$$

where R' = new shape factor
μ = a function expressible by

$$\mu^2 = \int \frac{\Phi^2\, dA}{AK^2} \quad (4.7)$$

where Φ = torsional function for the cross section
A = area of cross section of the bar
K = radius of gyration of the cross section of the bar

The approximate solution for μ^2 has been obtained in the case of rectangular cross sections with dimensions a and b. If $s = a/b$ is less than 3, the equation takes the form

$$\frac{\mu^2}{b^2} = \frac{1}{12(1+s^2)}\left(s^2 + \frac{4.577}{s}\tanh\frac{\pi s}{2} + 2.397 \tanh^2\frac{\pi s}{2} - 7.190\right) \quad (4.8)$$

for $s = 1$ (square cross section), this becomes

$$\frac{\mu^2}{b^2} = 0.000862 \quad (4.9)$$

Therefore, from Eqs. (4.6) and (4.9) we get the shape factor for the finite bar with square cross section as follows:

$$\frac{R'}{R} = 1 + 0.00851\left(\frac{nb}{l}\right)^2 \quad (4.10)$$

Tefft and Spinner[5] tested this equation experimentally by determining the shape factor from the resonant frequencies of uniform bars of steel by changing the ratio b/l. They found the agreement between theory and experiment to be quite satisfactory for lower values of the ratio b/l but unsatisfactory when the ratio b/l becomes large. The empirical relationship suggested by Tefft and Spinner based on their study is in the form

$$\frac{R'}{R} = \left[1 + n^2\left(\frac{b}{l}\right)^3 (0.01746 + 0.00148n + 0.00009n^2)\right] \quad (4.11)$$

Elastic Constants and Their Measurement

TABLE 4.1 Shape Factor R for Rectangular Cross Section

b/a	0	1	2	3	4
1.00	1.18559	1.18570	1.18604	1.18659	1.18735
1.10	1.19601	1.19809	1.20034	1.20276	1.20534
1.20	1.22402	1.22764	1.23139	1.23528	1.23930
1.30	1.26605	1.27093	1.27592	1.28102	1.28623
1.40	1.31975	1.32569	1.33173	1.33787	1.34411
1.50	1.38349	1.39037	1.39734	1.40439	1.41153
1.60	1.45613	1.46385	1.47165	1.47953	1.48749
1.70	1.53685	1.54534	1.55391	1.56255	1.57126
1.80	1.62503	1.63424	1.64352	1.65287	1.66229
1.90	1.72021	1.73010	1.74005	1.75008	1.76016
2.00	1.82203	1.83257	1.84317	1.85383	1.86456
2.10	1.93022	1.94138	1.95260	1.96389	1.97523
2.20	2.04455	2.05632	2.06814	2.08002	2.09197
2.30	2.16485	2.17721	2.18962	2.20209	2.21461
2.40	2.29098	2.30391	2.31689	2.32994	2.34304
2.50	2.42282	2.43631	2.44986	2.46347	2.47713
2.60	2.56026	2.57431	2.58842	2.60258	2.61680
2.70	2.70324	2.71784	2.73250	2.74721	2.76197
2.80	2.85169	2.86683	2.88203	2.89728	2.91258
2.90	3.00584	3.02122	3.03695	3.05274	3.06858
3.00	3.16475	3.18096	3.19723	3.21355	3.22993
3.10	3.32928	3.34603	3.36282	3.37967	3.39657
3.20	3.49909	3.51637	3.53369	3.55106	3.56849
3.30	3.67426	3.69196	3.70980	3.72770	3.74565
3.40	3.85425	3.87277	3.89114	3.90956	3.92803
3.50	4.03985	4.05879	4.07767	4.09661	4.11560
3.60	4.23063	4.24998	4.26939	4.28884	4.30835
3.70	4.42648	4.44635	4.46627	4.48624	4.50626
3.80	4.62748	4.64786	4.68829	4.68878	4.70932
3.90	4.83361	4.85451	4.87546	4.89645	4.91750
4.00	5.04487	5.06628	5.08774	5.10925	5.13081
4.10	5.26125	5.28317	5.30514	5.32716	5.34923
4.20	5.48273	5.50516	5.52764	5.55017	5.57275
4.30	5.70930	5.73224	5.75523	5.77827	5.80136
4.40	5.94097	5.96442	5.98791	6.01146	6.03506
4.50	6.17772	6.20167	6.22568	6.24973	6.27384
4.60	6.41954	6.44400	6.46852	6.49308	6.51769
4.70	6.66644	6.69140	6.71642	6.74149	6.76661
4.80	6.91839	6.94387	6.96939	6.99497	7.02060
4.90	7.17541	7.20139	7.22743	7.25351	7.27964
5.00	7.43749	7.46398	7.49051	7.51710	7.54374

5	6	7	8	9
1.18831	1.18948	1.19083	1.19238	1.19411
1.20808	1.21097	1.21401	1.21720	1.22054
1.24345	1.24773	1.25213	1.25665	1.26129
1.29156	1.29699	1.30252	1.30816	1.31390
1.35044	1.35687	1.36339	1.37000	1.37670
1.41876	1.42607	1.43346	1.44093	1.44849
1.49552	1.50364	1.51183	1.52009	1.52843
1.58004	1.58890	1.59782	1.60682	1.61589
1.67177	1.68132	1.69095	1.70063	1.71039
1.77031	1.78053	1.79081	1.80115	1.81156
1.87535	1.88620	1.89711	1.90809	1.91912
1.98663	1.99810	2.00962	2.02121	2.03285
2.10397	2.11603	2.12815	2.14032	2.15256
2.22720	2.23984	2.25254	2.26530	2.27811
2.35619	2.36940	2.38267	3.29600	2.40938
2.49084	2.50462	2.51845	2.53233	2.54627
2.63107	2.64539	2.65977	2.67421	2.68870
2.77679	2.79166	2.80658	2.82156	2.83600
2.92794	2.94335	2.95882	2.97434	2.98901
3.08448	3.10043	3.11643	3.13248	3.14859
3.24635	3.26283	3.27937	3.29595	3.31259
3.41353	3.43054	3.44760	3.46471	3.48188
3.58597	3.60351	3.62109	3.63873	3.65642
3.76366	3.78171	3.79982	3.81798	3.83619
3.94655	3.96513	3.98376	4.00244	4.02117
4.13465	4.15374	4.17288	4.19208	4.21133
4.32791	4.34752	4.36718	4.38690	4.40666
4.52634	4.54646	4.56664	4.58687	4.60715
4.72990	4.75054	4.77123	4.79198	4.81277
4.93860	4.95975	4.98096	5.00221	5.02352
5.15242	5.17409	5.19580	5.21756	5.23938
5.37135	5.39352	5.41575	5.43802	5.46035
5.59538	5.61806	5.64080	5.66358	5.68642
5.82450	5.84769	5.87094	5.89423	5.91758
6.05871	6.08241	6.10616	6.12996	6.15381
6.29800	6.32220	6.34646	6.37077	6.39513
6.54236	6.56707	6.59184	6.61665	6.64152
6.79178	6.81700	6.84228	6.86760	6.89297
7.04627	7.07200	7.09778	7.12361	7.14949
7.30582	7.33205	7.35834	7.38467	7.41106
7.57042	7.59716	7.62395	7.65079	7.67768

For the infinite bar with rectangular cross section, the solution by Davies or the one by Timoshenko[6] can be used. Timoshenko's equation may be expressed by the equation (Spinner and Tefft[2]) (where $b > a$)

$$R = \frac{1 + (b/a)^2}{4 - 2.521(a/b)\left(1 - \dfrac{1.991}{e^{\pi b/a} + 1}\right)} \qquad (4.12)$$

The Roark solution, quoted in the classical description of the dynamic resonance method by Pickett,[7] is in the form

$$R = \frac{1 + (b/a)^2}{4 - 2.52(a/b) + 0.21(a/b)^4} \qquad (4.13)$$

This is an approximation of Timoshenko's original equation based on a power series expansion. The table compiled by Hasselman[8] is based on Eq. (4.13), and gives the necessary correction factor R as a function of b/a from 1.000 to 10.999.

Spinner and Valore[9] tested these formulas experimentally and found that the difference between theory and experiment is within 0.2 percent if the ratio b/a is less than 2, but it increases to 1 percent as b/a becomes 4. Based on their results, Spinner and Tefft[2] recommend the use of the following empirical equation:

$$R = \left[\frac{1 + (b/a)^2}{4 - 2.521(a/b)\left(1 - \dfrac{1.991}{e^{\pi b/a} + 1}\right)}\right]\left(1 + \frac{0.00851 n^2 b^2}{l^2}\right)$$

$$- 0.060\left(\frac{nb}{l}\right)^{3/2}\left(\frac{b}{a} - 1\right)^2 \qquad (4.14)$$

The first bracket is the Timoshenko solution and the second one is the correction for the bar with finite length l. The calculation of Timoshenko's solution was made with a computer and the results are tabulated as a function of b/a in Table 4.1.

4.4 Flexural Vibration

As mentioned above, Young's modulus can be determined from either the flexural vibration or the longitudinal vibration. Generally speaking, it is easier to excite the flexural vibration than the longitudinal vibration, especially for thin specimens. For this reason, the flexural vibration is more practical and important for determining Young's modulus.

The general equation which relates Young's modulus and the flexural resonant frequency f_E is (Timoshenko,[6] Goens,[10] Pickett[7])

$$E = \left(\frac{2\pi l^2 f_E}{Km^2}\right)^2 \rho T \qquad (4.15)$$

where $K =$ radius of gyration of the cross section about the axis perpendicular to the plane of vibration
$m =$ constant depending on the mode of vibration
$T =$ shape factor, which depends upon the shape, size, and Poisson's ratio of the specimen and the mode of vibration

The value of K is known to be $d/4$ for a circular cross section and $a/\sqrt{12}$ for a rectangular cross section with the dimension a in the direction of vibration. The value of m is 4.7300 for the fundamental flexural vibration, 7.8532 for the first overtone, and 10.9956 for the second overtone. The difference between the shape factors T and R results from the fact that T depends upon Poisson's ratio as well as the shape and size of the specimen while R is independent of Poisson's ratio. Therefore, in order to calculate Young's modulus accurately from the flexural vibration, one must also know the value of Poisson's ratio. Since Poisson's ratio of a *homogeneous isotropic* body is expressed by

$$\sigma = \frac{E}{2G} - 1 \qquad (4.16)$$

one can calculate Poisson's ratio by determining both the shear modulus and Young's modulus. Therefore, one should obtain both torsional and flexural resonant frequencies even if only the value of Young's modulus is needed.

Pickett[7] showed that Goens'[10] solution of Timoshenko's differential equation for bars with free ends is in very close agreement with his result, derived by means of the mathematical theory of elasticity—if Poisson's ratio is known. So far as the shape factor T is concerned, the equation given by Goens is valid to at least three significant figures. Pickett's differential equations are in the form

$$\frac{E r_z^2}{\rho} \frac{\partial^4 v}{\partial x^4} + \frac{\partial^2 v}{\partial t^2} - r_z^2 \left(1 + \frac{E}{k'G}\right) \frac{\partial^4 v}{\partial x^2 \partial t^2} + \frac{r_z \rho^2}{k'G} \frac{\partial^4 v}{\partial t^4} = 0 \qquad (4.17)$$

where $r =$ radii of gyration of the cross section with respect to centroidal z and y axis
$v =$ displacement in y direction at time t
$x =$ coordinate in the direction of length
$k' =$ constant introduced by Timoshenko to account for the effect of shear on the slope of the elastic line

The first two terms form the basic equations for flexural vibration. The term $-r_z^2(\partial^4 v/\partial x^2 \partial t^2)$ is the one needed to correct for the effect of shear. A term $(r_z^2 - r_y^2)(\partial^4 v/\partial x^2 \partial t^2)$ may also be added to correct for the effect of lateral inertia, but, as discussed by Pickett,[7] this term is insignificant and, in fact, is zero for bars with square cross section. Therefore, when one deals with specimens of nearly square cross section, its effect can be neglected.

Starting from Eq. (4.17), Pickett[7] expressed the equations for determining the flexural resonance frequencies based upon the three-dimensional differential equations of elasticity. However, five complicated simultaneous equations are involved for a given length, diameter, Young's modulus, and Poisson's ratio, so that it is almost impossible to solve them without using a numerical method with the aid of a computer.

4.4.1 Cylindrical Rods. From Eq. (4.15) the relationship between Young's modulus and the fundamental flexural vibration for cylindrical rods is

$$E = 1.261886 \left(\frac{l^2 f_E}{d}\right)^2 \rho T_1 \tag{4.18}$$

and for the first overtone it is

$$E = 0.1660703 \left(\frac{l^2 f_E}{d}\right)^2 \rho T_2 \tag{4.19}$$

Tefft[11] performed the numerical calculations for T_1 and T_2 using Pickett's solution. His results are reproduced here as Tables 4.2 and 4.3. Experimental work by Spinner, Reichard, and Tefft[12] confirmed the earlier calculation made by Tefft, and they also compared the experimental data with the original values of Pickett, who gave an approximate calculation for some values of Poisson's ratio. They found that the value of T given by Pickett is too small; the difference between the values of Tefft and of Pickett is about 0.6 percent for a specimen with $d/l = 0.20$ and $\sigma = 0.33$.

4.4.2 Rectangular Bars For a rectangular cross section, the approximate equation has been derived by Pickett[7] and tested by Spinner, Reichard, and Tefft,[12] who found Pickett's formula to be in good agreement with experiment. The equation to be used is of the form

$$E = 0.94642 \left(\frac{l^2 f_E}{t}\right)^2 \rho T_3 \tag{4.20}$$

where t is the dimension of the cross section parallel to the direction of vibration, and T_3 is of the form

$$T_3 = 1 + 6.585(1 + 0.0752\sigma + 0.8109\sigma^2)\left(\frac{t}{l}\right)^2 - 0.868\left(\frac{t}{l}\right)^4$$

$$- \frac{8.340(1 + 0.2023\sigma + 2.173\sigma^2)(t/l)^4}{1 + 6.338(1 + 0.14081\sigma + 1.536\sigma^2)(t/l)^2} \tag{4.21}$$

The values of T_3 were calculated for different values of σ and (t/l) and the results are given in Table 4.4.

4.5 Longitudinal Vibration

If the specimen has the shape of a cylindrical rod or square bar, Young's modulus can be determined accurately from the longitudinal resonances of the specimen. This problem has been treated theoretically by several investigators, including Rayleigh,[13] Love,[14] and Lamb.[15] The differential equations for cylinders of infinite length were solved numerically by Bancroft,[16] and though not exact because of the finite specimen length, the solution is accurate in most instances. The equation has been tested by Tefft and Spinner.[5]

The equation which relates the longitudinal resonant frequency and Young's modulus is in the form

$$E = \frac{1}{U}\left(\frac{2lf_L}{n}\right)^2 \rho \quad (4.22)$$

where U is the correction factor which involves the shape and size of the specimen, wavelength, and Poisson's ratio. In the longitudinal resonance vibration of cylindrical rods, wavelength λ of standing waves is expressed by Eq. (4.1), and U for the case where $d/\lambda \ll 1$ is in the form (Rayleigh[13])

$$U \simeq 1 - \left(\frac{\pi n \sigma d}{\sqrt{8}l}\right)^2 \quad (4.23)$$

The experimental data of Tefft and Spinner[5] show that the value calculated from Eq. (4.23) is in good agreement within an accuracy of 0.1 percent. For specimens with a square cross section, d in Eq. (4.23) is replaced by $2b/\sqrt{3}$.

When more accurate values are desirable, then the empirical equation given by Tefft and Spinner may be suitable.
For cylinders

$$E = \frac{1}{U}\left(\frac{2lf_L}{n}\right)^2 \rho \left[1 + 0.0075\left(\frac{nd}{2l}\right)^2\right] \quad (4.24)$$

For square bars

$$E = \frac{1}{U}\left(\frac{2lf_L}{n}\right)^2 \rho \left[1 - 0.055\left(\frac{nt}{9l}\right)^4\right] \quad (4.25)$$

An appropriate equation for bars with a rectangular cross section has not been derived, and the use of the flexural vibration is therefore recommended.

TABLE 4.2 Correction Factor for the Fundamental Mode of Flexural Vibration of

Diameter-to-length ratio d/l	0.00	0.05	0.10	0.15	Poisson's 0.20
0.00	1.000000	1.000000	1.000000	1.000000	1.000000
0.02	1.001954	1.001979	1.002004	1.002029	1.002053
0.04	1.007804	1.007905	1.008004	1.008102	1.008199
0.06	1.017522	1.017748	1.017968	1.018186	1.018405
0.08	1.031064	1.031461	1.031848	1.032233	1.032618
0.10	1.048367	1.048983	1.049580	1.050174	1.050765
0.12	1.069357	1.070225	1.071077	1.071920	1.072753
0.14	1.093950	1.095111	1.096256	1.097378	1.098495
0.16	1.122061	1.123544	1.125007	1.126452	1.127884
0.18	1.153592	1.155430	1.157245	1.159039	1.160817
0.20	1.188458	1.190676	1.192869	1.195038	1.197191
0.22	1.226567	1.229189	1.231781	1.234351	1.236906
0.24	1.267868	1.270881	1.273894	1.276886	1.279865
0.26	1.312189	1.315668	1.319120	1.322555	1.325980
0.28	1.359546	1.363477	1.367381	1.371276	1.375167
0.30	1.409842	1.414233	1.418606	1.422974	1.427352
0.32	1.463014	1.467873	1.472724	1.477584	1.482465
0.34	1.519006	1.524340	1.529680	1.535043	1.540446
0.36	1.577766	1.583580	1.589417	1.595298	1.601240
0.38	1.639253	1.645548	1.651888	1.658300	1.664800
0.40	1.703422	1.710202	1.717054	1.724007	1.731082
0.42	1.770243	1.777505	1.784874	1.792382	1.800052
0.44	1.839681	1.847424	1.855317	1.863393	1.871677
0.46	1.911712	1.919934	1.928355	1.937012	1.945932
0.48	1.986311	1.995009	2.003964	2.013216	2.022795
0.50	2.063458	2.072628	2.082123	2.091985	2.102247
0.52	2.143136	2.152775	2.162814	2.173303	2.184276
0.54	2.225331	2.235431	2.246024	2.257157	2.268871
0.56	2.310030	2.320588	2.331739	2.343539	2.356026
0.58	2.397222	2.408232	2.419951	2.432439	2.445736
0.60	2.486902	2.498357	2.510654	2.523855	2.538002

Cylinders T_1 as a Function of the Diameter-to-Length Ratio d/l and Poisson's Ratio

ratio σ					
0.25	0.30	0.35	0.40	0.45	0.50
1.000000	1.000000	1.000000	1.000000	1.000000	1.000000
1.002077	1.002100	1.002124	1.002147	1.002170	1.002193
1.008295	1.008388	1.008482	1.008575	1.008666	1.008757
1.018619	1.018826	1.019038	1.019245	1.019450	1.019653
1.032994	1.033360	1.033733	1.034096	1.034459	1.034818
1.051344	1.051916	1.052484	1.053050	1.053610	1.054170
1.073577	1.074393	1.075202	1.076008	1.076808	1.077604
1.099599	1.100694	1.101782	1.102864	1.103941	1.105015
1.129302	1.130711	1.132113	1.133509	1.134901	1.136288
1.162585	1.164337	1.166086	1.167830	1.169569	1.171308
1.199332	1.201464	1.203591	1.205714	1.207838	1.209961
1.239449	1.241986	1.244521	1.247057	1.249596	1.252141
1.282836	1.285807	1.288779	1.291757	1.294746	1.297747
1.329403	1.332832	1.336270	1.339723	1.343194	1.346687
1.379065	1.382977	1.386911	1.390869	1.394857	1.398879
1.431747	1.436169	1.440625	1.445121	1.449661	1.454250
1.487380	1.492337	1.497345	1.502411	1.507540	1.512737
1.545901	1.551420	1.557012	1.562682	1.568439	1.574286
1.607259	1.613367	1.619572	1.625884	1.632309	1.638853
1.671405	1.678129	1.684983	1.691975	1.699113	1.706401
1.738298	1.745669	1.753207	1.760921	1.768818	1.776904
1.807904	1.815954	1.824214	1.832694	1.841401	1.850342
1.880193	1.888956	1.897979	1.907273	1.916846	1.926702
1.955140	1.964653	1.974485	1.984646	1.995143	2.005980
2.032727	2.043030	2.053718	2.064802	2.076288	2.088178
2.112937	2.124075	2.135673	2.147741	2.160285	2.173305
2.195762	2.207781	2.220345	2.233465	2.247142	2.261377
2.281194	2.294146	2.307741	2.321984	2.336876	2.352415
2.369231	2.383174	2.397866	2.413311	2.429508	2.446448
2.459873	2.474869	2.490733	2.507467	2.525064	2.543511
2.553126	2.569244	2.586362	2.604478	2.623578	2.643647

TABLE 4.3 Correction Factor for the First Overtone of Flexural Vibration of Cyl-

Diameter-to-length ratio d/l	Poisson's				
	0.00	0.05	0.10	0.15	0.20
0.00	1.000000	1.000000	1.000000	1.000000	1.000000
0.01	1.001352	1.001376	1.001399	1.001422	1.001445
0.02	1.005404	1.005499	1.005592	1.005683	1.005774
0.03	1.012144	1.012355	1.012564	1.012770	1.012973
0.04	1.021552	1.021926	1.022295	1.022660	1.023019
0.05	1.033605	1.034188	1.034760	1.035326	1.035885
0.06	1.048267	1.049105	1.049924	1.050734	1.051537
0.07	1.065511	1.066641	1.067750	1.068844	1.069926
0.08	1.085296	1.086759	1.088198	1.089613	1.091015
0.09	1.107583	1.109415	1.111219	1.112996	1.114756
0.10	1.132330	1.134568	1.136772	1.138949	1.141103
0.11	1.159496	1.162177	1.164820	1.167429	1.170012
0.12	1.189048	1.192205	1.195318	1.198394	1.201441
0.13	1.220947	1.224613	1.228230	1.231807	1.235353
0.14	1.255166	1.259371	1.263523	1.267633	1.271711
0.15	1.291674	1.296449	1.301168	1.305844	1.310487
0.16	1.330448	1.335824	1.341140	1.346414	1.351656
0.17	1.371471	1.377477	1.383422	1.389325	1.395199
0.18	1.414728	1.421393	1.427999	1.434563	1.441104
0.19	1.460212	1.467565	1.474861	1.482122	1.489364
0.20	1.507918	1.515991	1.524010	1.532000	1.539980
0.21	1.557849	1.566672	1.575448	1.584202	1.592958
0.22	1.610014	1.619619	1.629185	1.638741	1.648312
0.23	1.664427	1.674847	1.685239	1.695636	1.706063
0.24	1.721106	1.732377	1.743634	1.754912	1.766240
0.25	1.780080	1.792238	1.804400	1.816603	1.828880
0.26	1.841381	1.854465	1.867575	1.880751	1.894027
0.27	1.905048	1.919102	1.933207	1.947405	1.961734
0.28	1.971129	1.986199	2.001348	2.016625	2.032066
0.29	2.039679	2.055815	2.072063	2.088477	2.105095
0.30	2.110759	2.128016	2.145425	2.163041	2.180906
0.31	2.184442	2.202881	2.221516	2.240405	2.259595
0.32	2.260809	2.280496	2.300430	2.230672	2.341270
0.33	2.339949	2.360959	2.382273	2.403955	2.426056
0.34	2.421965	2.444380	2.467163	2.490383	2.514090
0.35	2.506968	2.530881	2.555234	2.580098	2.605527
0.36	2.595085	2.620598	2.646633	2.673262	2.700540

inders T_2 as a Function of the Diameter-to-Length Ratio d/l and Poisson's Ratio

ratio σ					
0.25	0.30	0.35	0.40	0.45	0.50
1.000000	1.000000	1.000000	1.000000	1.000000	1.000000
1.001467	1.001489	1.001511	1.001533	1.001554	1.001576
1.005863	1.005952	1.006039	1.006126	1.006212	1.006297
1.013174	1.013373	1.013569	1.013763	1.013957	1.014149
1.023376	1.023728	1.024077	1.024421	1.024763	1.025103
1.036440	1.036987	1.037530	1.038064	1.038597	1.039128
1.052327	1.053108	1.053883	1.054653	1.055417	1.056178
1.070995	1.072054	1.073103	1.074146	1.075181	1.076211
1.092401	1.093775	1.095137	1.096491	1.097837	1.099177
1.116497	1.118223	1.119937	1.121641	1.123337	1.125025
1.143235	1.145352	1.147455	1.149547	1.151630	1.153705
1.172573	1.175115	1.177643	1.180161	1.182668	1.185170
1.204465	1.207470	1.210461	1.213439	1.216410	1.219376
1.238874	1.242376	1.245865	1.249343	1.252816	1.256285
1.275764	1.279800	1.283822	1.287838	1.291851	1.295863
1.315106	1.319710	1.324305	1.328896	1.333488	1.338084
1.356876	1.362084	1.367288	1.372493	1.377707	1.382928
1.401055	1.406905	1.412756	1.418616	1.424490	1.430382
1.447633	1.454162	1.460700	1.467255	1.473834	1.480442
1.496602	1.503850	1.511116	1.518411	1.525741	1.533111
1.547966	1.555973	1.564010	1.572090	1.580217	1.588400
1.601732	1.610540	1.619395	1.628306	1.637282	1.646330
1.657917	1.667572	1.677290	1.687084	1.696961	1.706930
1.716543	1.727092	1.737726	1.748455	1.759290	1.770238
1.777642	1.789136	1.800738	1.812460	1.824312	1.836303
1.841254	1.853748	1.866375	1.879151	1.892085	1.905185
1.907428	1.920979	1.934694	1.948588	1.962672	1.976953
1.976222	1.990892	2.005761	2.020844	2.036152	2.051692
2.047703	2.063560	2.079655	2.096003	2.112614	2.129495
2.121950	2.139068	2.156467	2.174161	2.192161	2.210474
2.199055	2.217513	2.236300	2.255430	2.274913	2.294753
2.279119	2.299006	2.319274	2.339937	1.261003	2.382477
2.362262	2.383673	2.405523	2.427824	2.450585	2.473806
2.448613	2.471654	2.495197	2.519253	2.543828	2.568923
2.538324	2.563111	2.588468	2.614407	2.640929	2.668033
2.631559	2.658220	2.685528	2.713488	2.742103	2.771366
2.728506	2.757184	2.786589	2.816726	2.847593	2.879180

TABLE 4.4 Shape Factor T_3 As a Function of t/l and σ

t/l	Poisson's ratio σ								
	0.05	0.10	0.15	0.20	0.25	0.30	0.35	0.40	0.45
0.00000	00000	1.00000	1.00000	1.00000	1.00000	1.00000	1.00000	1.00000	1.00000
0.00500	1.00016	1.00016	1.00016	1.00017	1.00017	1.00018	1.00018	1.00019	1.00019
0.01000	1.00066	1.00066	1.00067	1.00068	1.00070	1.00072	1.00074	1.00076	1.00078
0.01500	1.00148	1.00150	1.00152	1.00155	1.00158	1.00162	1.00166	1.00171	1.00177
0.02000	1.00264	1.00267	1.00271	1.00275	1.00281	1.00288	1.00296	1.00305	1.00315
0.02500	1.00413	1.00417	1.00423	1.00430	1.00439	1.00450	1.00462	1.00476	1.00492
0.03000	1.00595	1.00601	1.00609	1.00619	1.00632	1.00648	1.00666	1.00686	1.00708
0.03500	1.00809	1.00817	1.00829	1.00843	1.00861	1.00882	1.00906	1.00933	1.00964
0.04000	1.01057	1.01067	1.01082	1.01101	1.01124	1.01151	1.01182	1.01218	1.01258
0.04500	1.01337	1.01350	1.01368	1.01392	1.01421	1.01456	1.01496	1.01541	1.01592
0.05000	1.01650	1.01666	1.01688	1.01718	1.01754	1.01796	1.01845	1.01901	1.01963
0.05500	1.01995	1.02014	1.02041	1.02077	1.02120	1.02172	1.02231	1.02298	1.02374
0.06000	1.02372	1.02395	1.02428	1.02470	1.02521	1.02582	1.02653	1.02733	1.02822
0.06500	1.02781	1.02808	1.02847	1.02896	1.02956	1.03028	1.03110	1.03204	1.03309
0.07000	1.03223	1.03254	1.03298	1.03355	1.03425	1.03508	1.03604	1.03712	1.03834
0.07500	1.03696	1.03732	1.03783	1.03848	1.03928	1.04023	1.04132	1.04257	1.04396
0.08000	1.04201	1.04242	1.04299	1.04374	1.04464	1.04572	1.04696	1.04838	1.04995
0.08500	1.04738	1.04784	1.04848	1.04932	1.05034	1.05155	1.05295	1.05454	1.05632
0.09000	1.05306	1.05357	1.05429	1.05522	1.05637	1.05772	1.05929	1.06107	1.06306
0.09500	1.05905	1.05961	1.06042	1.06145	1.06273	1.06423	1.06597	1.06795	1.07016
0.10000	1.06534	1.06597	1.06686	1.06800	1.06941	1.07107	1.07300	1.07518	1.07763
0.10500	1.07195	1.07264	1.07361	1.07487	1.07642	1.07825	1.08036	1.08275	1.08545
0.11000	1.07886	1.07961	1.08068	1.08206	1.08375	1.08575	1.08807	1.09069	1.09364
0.11500	1.08607	1.08689	1.08805	1.08956	1.09140	1.09358	1.09611	1.09897	1.10217
0.12000	1.09358	1.09447	1.09574	1.09737	1.09937	1.10174	1.10448	1.10759	1.11107
0.12500	1.10189	1.10236	1.10372	1.10549	1.10765	1.11021	1.11318	1.11654	1.12031
0.13000	1.10949	1.11054	1.11201	1.11391	1.11625	1.11901	1.12221	1.12583	1.12989
0.13500	1.11789	1.11901	1.12060	1.12264	1.12515	1.12812	1.13156	1.13546	1.13983
0.14000	1.12658	1.12778	1.12948	1.13167	1.13436	1.13755	1.14123	1.14542	1.15010
0.14500	1.13556	1.13685	1.13866	1.14100	1.14388	1.14729	1.15123	1.15570	1.16071
0.15000	1.14483	1.14620	1.14813	1.15063	1.15370	1.15733	1.16154	1.16631	1.17166
0.15500	1.15438	1.15584	1.15789	1.16055	1.16382	1.16769	1.17216	1.17724	1.18294
0.16000	1.16421	1.16576	1.16794	1.17077	1.17424	1.17835	1.18310	1.18850	1.19455
0.16500	1.17422	1.17596	1.17828	1.18127	1.18495	1.18930	1.19434	1.20007	1.20648
0.17000	1.18471	1.18644	1.18890	1.19207	1.19595	1.20056	1.20590	1.21196	1.21875
0.17500	1.19537	1.19721	1.19979	1.20314	1.20725	1.21212	1.21775	1.22416	1.23133
0.18000	1.20631	1.20824	1.21097	1.21450	1.21883	1.22397	1.22991	1.23667	1.24424
0.18500	1.21752	1.21955	1.22243	1.22614	1.23070	1.23611	1.24237	1.24949	1.25746
0.19000	1.22899	1.23113	1.23415	1.23806	1.24286	1.24854	1.25513	1.26261	1.27100
0.19500	1.24074	1.24298	1.24615	1.25025	1.25529	1.26126	1.26818	1.27604	1.28486
0.20000	1.25274	1.25510	1.25842	1.26272	1.26800	1.27427	1.28152	1.28977	1.29902
0.20500	1.26501	1.26748	1.27546	1.27546	1.28100	1.28756	1.29516	1.30380	1.31349
0.21000	1.27754	1.28012	1.28376	1.28848	1.29426	1.30113	1.30908	1.31813	1.32828
0.21500	1.29033	1.29302	1.29683	1.30175	1.30780	1.31498	1.32329	1.33275	1.34336
0.22000	1.30338	1.30619	1.31016	1.31530	1.32161	1.32911	1.33779	1.34767	1.35875
0.22500	1.31667	1.31960	1.32375	1.32911	1.33569	1.34351	1.35257	1.36288	1.37445
0.23000	1.33623	1.33328	1.33759	1.34318	1.35004	1.35819	1.36763	1.37838	1.39044
0.23500	1.34403	1.34721	1.35170	1.35751	1.36465	1.37314	1.38297	1.39416	1.40673
0.24000	1.35808	1.36138	1.36605	1.37210	1.37953	1.38836	1.39859	1.41024	1.42332
0.24500	1.37238	1.37581	1.38067	1.38695	1.39467	1.40385	1.41448	1.42660	1.44020

TABLE 4.4 Shape Factor T_3 as a Function of t/l and σ (*Continued*)

	Poisson's ratio σ								
t/l	0.05	0.10	0.15	0.20	0.25	0.30	0.35	0.40	0.45
0.25000	1.38692	1.39049	1.39553	1.40205	1.41007	1.41960	1.43065	1.44324	1.45738
0.25500	1.40171	1.40541	1.41064	1.41741	1.42573	1.43562	1.44710	1.46016	1.47485
0.26000	1.41674	1.42057	1.42600	1.43302	1.44165	1.45191	1.46381	1.47737	1.49261
0.26500	1.43201	1.43598	1.44160	1.44888	1.45782	1.46846	1.48079	1.49485	1.51066
0.27000	1.44753	1.45164	1.45745	1.46498	1.47425	1.48526	1.49805	1.51262	1.52900
0.27500	1.46327	1.46753	1.47354	1.48134	1.49093	1.50233	1.51557	1.53066	1.54762
0.28000	1.47926	1.48366	1.48988	1.49794	1.50786	1.51966	1.53335	1.54897	1.56653
0.28500	1.49548	1.50002	1.50645	1.51479	1.52504	1.53724	1.55141	1.56756	1.58573
0.29000	1.51193	1.51662	1.52327	1.53188	1.54248	1.55508	1.56972	1.58642	1.60521
0.29500	1.52862	1.53346	1.54032	1.54921	1.56015	1.57317	1.58830	1.60555	1.62497
0.30000	1.54553	1.55053	1.55761	1.56678	1.57808	1.59152	1.60714	1.62496	1.64501
0.30500	1.56268	1.56783	1.57513	1.58460	1.59625	1.61012	1.62624	1.64463	1.66533
0.31000	1.58005	1.58536	1.59289	1.60265	1.61466	1.62897	1.64559	1.66457	1.68594
0.31500	1.59765	1.60312	1.61088	1.62094	1.63332	1.64807	1.66521	1.68478	1.70682
0.32000	1.61547	1.62111	1.62910	1.63946	1.65222	1.66742	1.68508	1.70526	1.72797
0.32500	1.63352	1.63933	1.64755	1.65822	1.67136	1.68701	1.70521	1.72600	1.74941
0.33000	1.65179	1.65777	1.66623	1.67721	1.69074	1.70685	1.72559	1.74700	1.77112
0.33500	1.67029	1.67643	1.68514	1.69644	1.71036	1.72694	1.74623	1.76827	1.79310
0.34000	1.68900	1.69532	1.70427	1.71589	1.73021	1.74727	1.76712	1.78980	1.81535
0.34500	1.70794	1.71443	1.72364	1.73558	1.75030	1.76785	1.78826	1.81159	1.83789
0.35000	1.72709	1.73376	1.74322	1.75550	1.77063	1.78867	1.80965	1.83365	1.86069
0.35500	1.74646	1.75331	1.76303	1.77564	1.79119	1.80972	1.83130	1.85596	1.88376
0.36000	1.76605	1.77309	1.78306	1.79601	1.81198	1.83102	1.85319	1.87853	1.90711
0.36500	1.78585	1.79307	1.80331	1.81661	1.83301	1.85256	1.87533	1.90136	1.93072
0.37000	1.80587	1.81328	1.82379	1.83743	1.85427	1.87434	1.89771	1.92445	1.95460
0.37500	1.82610	1.83370	1.84448	1.85848	1.87575	1.89635	1.92035	1.94779	1.97875
0.38000	1.84654	1.85433	1.86539	1.87975	1.89747	1.91861	1.94322	1.97139	2.00317
0.38500	1.86719	1.87518	1.88652	1.90124	1.91941	1.94109	1.96635	1.99524	2.02785
0.39000	1.88805	1.89624	1.90786	1.92296	1.94159	1.96381	1.98971	2.01935	2.05279
0.39500	1.90912	1.91752	1.92942	1.94489	1.96399	1.98677	2.01322	2.04371	2.07800
0.40000	1.93040	1.93900	1.95120	1.96704	1.98661	2.00996	2.03717	2.06832	2.10348
0.40500	1.95189	1.96069	1.97318	1.98942	2.00946	2.03338	2.06126	2.09318	2.12921
0.41000	1.97358	1.98259	1.99538	2.01201	2.03253	2.05703	2.08559	2.11829	2.15521
0.41500	1.99548	2.00470	2.01779	2.03481	2.05582	2.08091	2.11016	2.14365	2.18147
0.42000	2.01758	2.02702	2.04042	2.05783	2.07934	2.10502	2.13497	2.16926	2.20799
0.42500	2.03988	2.04954	2.06325	2.08107	2.10308	2.12936	2.16001	2.19512	2.23477
0.43000	2.06239	2.07227	2.08628	2.10451	2.12703	2.15393	2.18529	2.22122	2.26180
0.43500	2.08510	2.09519	2.10953	2.12817	2.15121	2.17872	2.21081	2.24757	2.28910
0.44000	2.10800	2.11833	2.13298	2.15205	2.17560	2.20374	2.23656	2.27416	2.31665
0.44500	2.13110	2.14166	2.15664	2.17613	2.20021	2.22898	2.26254	2.30100	2.34445
0.45000	2.15441	2.16519	2.18050	2.20042	2.22503	2.25444	2.28876	2.32808	2.37251
0.45500	2.17791	2.18892	2.20457	2.22492	2.25007	2.28013	2.31520	2.35540	2.40082
0.46000	2.20160	2.21285	2.22883	2.24963	2.27533	2.30604	2.34188	2.38296	2.42939
0.46500	2.22549	2.23698	2.25330	2.27454	2.30079	2.33217	2.36879	2.41076	2.45821
0.47000	2.24957	2.26130	2.27797	2.29966	2.32647	2.35852	2.39592	2.43880	2.48727
0.47500	2.27384	2.28582	2.30284	2.32489	2.35236	2.38508	2.42328	2.46708	2.51659
0.48000	2.29831	2.31054	2.32790	2.35051	2.37845	2.41187	2.45087	2.49560	2.54616
0.48500	2.32296	2.33544	2.35317	2.37623	2.40476	2.43887	2.47869	2.52435	2.57597
0.49000	2.34781	2.36054	2.37862	2.40216	2.43127	2.46608	2.50673	2.55333	2.60603
0.49500	2.37284	2.38583	2.40428	2.42829	2.45799	2.49351	2.53499	2.58255	2.63634
0.50000	2.39806	2.41130	2.43012	2.45462	2.48492	2.52115	2.56347	2.61200	2.66689

4.6 Measuring System

A complete measuring system capable of determining the natural resonant frequencies of bars or rods is available from several companies. Such a system has a measuring range from 600 to 26,000 Hz, and provides for the measurement of internal friction. Alternatively it can be constructed using standard off-the-shelf components.

The system consists of (1) a driving circuit, (2) a pickup circuit, and (3) specimen support. Those who wish to build their own system can do so easily be following the excellent description of Spinner and Tefft.[2] A brief outline of their apparatus follows (see Fig. 4.1).

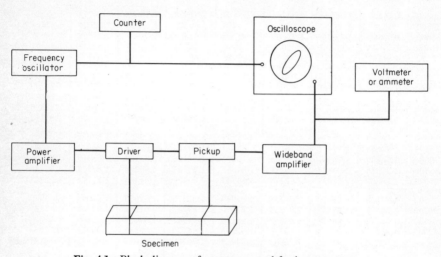

Fig. 4.1 Block diagram of apparatus used for bar resonance.

A stable, variable-frequency oscillator (to 50 kHz), provided with an automatic scanning system that reduces labor in detecting several resonant conditions, serves as the energy source. The frequency counter is used for the accurate determination of resonant frequencies, and so it should have a counting accuracy of 1 Hz on the 1-s interval (or 0.1 Hz on the 10-s interval) up to 100 kHz, By counting the frequency in 10-s intervals, the frequency may be determined with five significant figures when the resonance frequency is below 10 kHz. If the counter has a period counting mode, it is preferable to use this mode rather than the frequency counting mode because of increased resolution of low-frequency signals in short times. The power amplifier should have a maximum output of at least 20 W with a flat frequency response over the entire range used. This is especially important if one wants to determine the internal friction of the specimen by measuring the half-width of the resonant spectrum. As the driver, a record-cutting head or a tweeter-type speaker can be used. However, because of the development of new types of high-power piezoelectric transducers, the commercially available units utilizing these new transducers are recommended. Excitation methods other than the standard resonance type will be described later.

The pickup transducer should be a sensitive one. A phonograph cartridge of the crystal or magnetic reluctance type was used by Spinner and Tefft.[2] The piezoelectric type, such as Rochelle salt, is incorporated in commercial units. The wideband pickup amplifier should have a flat frequency response comparable to the power amplifier. An oscilloscope with both x and y inputs is needed to obtain Lissajous patterns on the screen. A voltmeter or a milliammeter is useful in detecting the resonant condition since it gives an amplitude maximum at the pickup.

Specimen support, which provides the coupling of (1) energy from the driver to the specimen and (2) detecting the vibrations of the specimen with the pickup, is the crucial element in determining the elastic moduli of a rod or cylinder. Since a mechanical restraint of any sort applied to a specimen may affect its natural resonant frequencies, thus causing differing values of elastic moduli for the specimen to be realized, one must try to minimize any mechanical restraint.

A widely adopted method is to support the specimen on knife edges or foam rubber at the nodal points and excite it through flexible wires extending from the transducer or directly with an air column by means of a tweeter, as shown in Fig. 4.2. The nodes of the vibrating specimen are the positions of zero displacement in the direction of vibration. This depends upon the mode of vibration, as given in Table 4.5. However, best results have been observed on suspending the specimen from threads, one thread attached directly to the driver and the other to the pickup. To obtain accurate results, the position of the threads should be close to the nodes. When the position of the thread is far from the nodes, a frequency

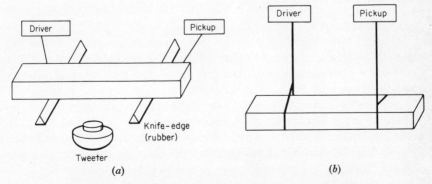

Fig. 4.2 Illustration of methods of coupling acoustic energy.

TABLE 4.5 Position of the Nodes Expressed as a Fraction of the Length of the Specimen

Mode of Vibration	Type of vibration		
	Flexural	Torsional	Longitudinal
Fundamental	0.224	0.500	0.500
	0.776		
First overtone	0.132	0.250	0.250
	0.500	0.750	0.750
	0.868		

shift of a few cycles may occur. The thread may be cotton, nylon, glass fiber, or even thin wire of a refractory metal for use at elevated temperatures. If the specimen is suspended in the manner shown in Fig. 4.2b, one can obtain both torsional vibration and flexural vibration measurements.

4.7 Identification of the Vibrating Mode

As described previously, there are three types of vibrations normally excited, namely, flexural, torsional, and longitudinal. Besides, a rectangular bar has two types for any mode of flexural vibration, depending upon the relation of the direction of vibration to either cross-sectional dimension. If the vibration is parallel to the short side of the cross section, it is generally called "flatwise" flexural vibration; and if it is parallel to the long side of the cross section, it is called "edgewise" flexural vibration. In order to calculate the elastic moduli of the specimen, one must know exactly the type and mode of the vibration.

As the scanning frequency approaches the resonant frequency of the specimen, a Lissajous pattern appears on the oscilloscope. The frequency at which maximum signal amplitude is observed is the resonant frequency. Since the Lissajous pattern, whose shape is controlled by the phase shift of the electronic signal, is an exact reflection of the mechanical vibration, it is possible to determine whether the motion of the specimen at the position of the driver and the pickup is of the same phase or opposite. When the Lissajous pattern is adjusted so that the direction of the diagonal is down to the left for the same phase, there will be no pattern at the nodes and a down-to-the-right direction will occur for the opposite phase, as shown in Fig. 4.3. This search can be done by either moving the position of the pickup or driver separately or simultaneously, or by traversing the side of the specimen with the pickup, without changing any other condition. By counting the number of nodes along the bar, one can determine the mode of vibration. Torsional modes give the change in pattern along the cross-sectional face of the bar, as shown in Fig. 4.3. Except for some

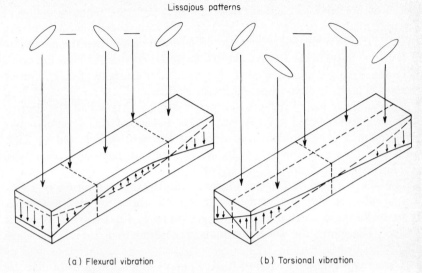

Fig. 4.3 The resonant vibrations in a rectangular bar.

specimens which have an unusually low shear modulus, the fundamental torsional resonant frequency for thin rods or bars is higher than the fundamental flexural resonant frequency. The fundamental flexural vibration has two nodes. The fundamental torsional vibration has only one node at the center. By suspending the specimen as shown in Fig. 4.2, the longitudinal vibration is almost impossible to excite. Therefore, if one observes a resonant condition which has a node at the center, after observing the resonant condition with two nodes, one can anticipate that it is the torsional vibration.

An alternative approach is to identify the mode of vibration from the position of the nodes. As shown in Table 4.5, torsional and longitudinal vibrations give the same nodal position: the fundamental having a node at the center, and the first overtone at $0.25l$ and $0.75l$ from one end of the specimen. Flexural vibrations give slightly different nodal positions, which are $0.224l$ and $0.776l$ for the fundamental mode. These nodal points can be made visible by placing a fine substance such as SiC powder on the specimen, as suggested by Hasselman;[8] the grains will then accumulate at the position of the node.

4.8 Calculation of Elastic Moduli from Resonant Frequencies

Since the equation which relates Young's modulus with the flexural resonant frequency requires a knowledge of Poisson's ratio (σ), one starts the calculation by assuming a reasonable value for that ratio, say 0.25. From the shear modulus calculated from Eq. (4.3), and the tentative value of Young's modulus (using $\sigma = 0.25$), Poisson's ratio is calculated from Eq. (4.16). If this value disagrees with the assumed value, then the new value of Poisson's ratio is used to recalculate E until the values of both Poisson's ratio and Young's modulus each converge. An example of this iterative method is given here to illustrate the procedure.

A rectangular bar of hot-pressed polycrystalline MgO had the dimensions of $1.644 \times 0.977 \times 9.154$ cm. The density of this specimen was 3.574 g/cm^3. The fundamental resonant frequencies detected at 25°C were as follows: torsional $f_G = 26,681$ Hz; flatwise flexural $f_{E1} = 10,716$ Hz; and edgewise flexural $f_{E2} = 16,988$ Hz.

The shape factor R for the torsional fundamental vibration is 1.522, which value is obtained using Table 4.1 for the ratio of $b/a = 1.683$. If we make the correction for length according to Eq. (4.14), R becomes 1.520. Therefore, the shear modulus, according to Eq. (4.3), is 1,296 kbar.

Assuming a Poisson's ratio of 0.25, Young's modulus from the flatwise flexural resonant frequency is 3,083 kbar by Eq. (4.20), using a value of T_3 of 1.0790 for $t/l = 0.1067$. Poisson's ratio is, therefore, $(3,083/2 \times 1,296) - 1 = 0.189$. Since the difference between the assumed value of 0.25 and the determined value of 0.189 is large, the new value of T_3 should be used. The new value of T_3 for the same t/l but $\sigma = 0.19$ is 1.0771, yielding a Young's modulus of 3,078 kbar and a Poisson's ratio of 0.188, which now satisfactorily agrees with the value used (0.19). The same procedure was taken for the edgewise flexural vibration. By using 1.2129 as the value of T_3 for $t/l = 0.1796$ and $\sigma = 0.19$, Young's modulus was found to be 3,076 kbar and Poisson's ratio 0.187.

4.9 The Effect of Orientation upon the Relation of Young's Modulus and Shear Modulus to the Elastic Compliances

The elastic moduli obtained with the resonance method give information about the elastic compliances along the long axis of the bar. In an elastically isotropic body, such as a well-prepared polycrystalline specimen or glass, the elastic moduli are identical in any direction. In an anisotropic body the elastic moduli depend on the orientation of the specimen. The theory relating the elastic moduli and compliances in terms of specimen orientation has been worked out by Love,[14] Voigt,[17] Goens,[18] and Brown.[19] A summary of their works can be found in the excellent review papers by Hearmon.[20,21]

Because of the existence of elastic compliances which relate an extensional stress to a shear strain, an *anisotropic* specimen subjected to a pure extension undergoes both extensional and shear strains. Consequently, the specimen twists and bends at the same time even though only a bending moment or a twisting couple is applied. Therefore, a distinction is necessary between "free" and "pure" elastic moduli. The free elastic modulus corresponds to the condition where the applied stress (a bending moment for Young's modulus and a twisting couple for the shear modulus) results in a combination of flexes and torsions, while the pure elastic modulus corresponds to the condition where the distortion is purely flexural or torsional, depending on which type of stress is applied. For isotropic materials there is no distinction between free and pure elastic moduli. For anisotropic materials, the free elastic moduli are not the pure elastic moduli unless the long axis of the bar specimen is oriented along one of its major crystallographic axes. The pure elastic moduli can be calculated, however, from the measured free elastic moduli by knowing the exact crystal orientation of the specimen and the relations between the free and pure elastic moduli. These relations have been given by Goens,[18] Brown,[19] and Hearmon.[20,21]

As described before, Young's modulus is defined as the ratio tensile stress/linear strain and the shear modulus as the ratio shear stress/shear strain. Thus, from linear equations of Hooke's law expressing the strain components in terms of the stress components (see Chap. 2) one can obtain the following relationships if one employs a coordinate system where the z' axis is along the rod axis and the other rectangular coordinates are x' and y'.

$$E_f = \frac{1}{s'_{33}} \tag{4.26}$$

$$G_f = \frac{2}{(s'_{44} + s'_{55})} \tag{4.27}$$

where the primes indicate that the elastic compliances s_{ij} are referred to the system x', y', z' (not necessarily coincident with the crystallographic axis).

The difference between free and pure elastic moduli is (Hearmon[21])

$$E_p = \frac{E_f}{(1-\varepsilon)} \tag{4.28}$$

$$G_p = \frac{G_f}{(1-\varepsilon)} \tag{4.29}$$

where ε is given by

$$\varepsilon = \frac{(s'^2_{34} + s'^2_{35})}{s'_{33}(s'_{44} + s'_{55})} \tag{4.30}$$

For simplest application of the results obtained, it is desirable to choose an orientation for which $\varepsilon = 0$ whenever possible.

The free moduli can be expressed in terms of the unprimed elastic compliances by a tensor transformation. Suppose the new coordinate system x_1', x_2', x_3' is related to the standard rectangular coordinate system x_1, x_2, x_3, by

$$x_1' = \alpha_1 x_1 + \alpha_2 x_2 + \alpha_3 x_3$$
$$x_2' = \beta_1 x_1 + \beta_2 x_2 + \beta_3 x_3 \quad (4.31)$$
$$x_3' = \gamma_1 x_1 + \gamma_2 x_2 + \gamma_3 x_3$$

where the α, β, and γ are the elements of the transformation (see Chap. 2). In this more compact notation, we recognize that $\alpha_1 = a_{11}$, $\alpha_2 = a_{12}$, $\alpha_3 = a_{13}$; $\beta_1 = a_{21}$, $\beta_2 = a_{22}$, $\beta_3 = a_{23}$; $\gamma_1 = a_{31}$, $\gamma_2 = a_{32}$, $\gamma_3 = a_{33}$. This is achieved by building part of the subscript notation into the element notation, i.e., $\alpha = a_{1j}$, $\beta = a_{2j}$, $\gamma = a_{3j}$, $j = 1, 2, 3$. Since both systems are rectangular, the following relationships hold among the direction cosines:

$$\alpha_1^2 + \alpha_2^2 + \alpha_3^2 = 1$$
$$\beta_1^2 + \beta_2^2 + \beta_3^2 = 1$$
$$\gamma_1^2 + \gamma_2^2 + \gamma_3^2 = 1$$
$$\alpha_1^2 + \beta_1^2 + \gamma_1^2 = 1$$
$$\alpha_2^2 + \beta_2^2 + \gamma_2^2 = 1$$
$$\alpha_3^2 + \beta_3^2 + \gamma_3^2 = 1 \quad (4.32)$$
$$\alpha_1 \alpha_2 + \beta_1 \beta_2 + \gamma_1 \gamma_2 = 0$$
$$\alpha_2 \alpha_3 + \beta_2 \beta_3 + \gamma_2 \gamma_3 = 0$$
$$\alpha_3 \alpha_1 = \beta_3 \beta_1 + \gamma_3 \gamma_1 = 0$$
$$\alpha_1 \beta_1 + \alpha_2 \beta_2 + \alpha_3 \beta_3 = 0$$
$$\beta_1 \gamma_1 + \beta_2 \gamma_2 + \beta_3 \gamma_3 = 0$$
$$\gamma_1 \alpha_1 + \gamma_2 \alpha_2 + \gamma_3 \alpha_3 = 0$$

The transformation of the elastic compliances s_{33}', s_{44}', and s_{55}' can be made in terms of these direction cosines. The results obtained by Voigt[17] are given in Table 4.6. It is to be read downward in such a way that

$$s_{33}' = \gamma_1^4 s_{11} + 2\gamma_1^2 \gamma_2^2 s_{12} + 2\gamma_1^2 \gamma_3^2 s_{13} + \cdots \quad (4.33)$$

In the cubic system, the relationship between the s_{ij}' and s_{ij} is simplified because of the effect of cubic symmetry operations upon the compliance matrix (see Chap. 2). This results in the reduction of these compliances to $s_{11} = s_{22} = s_{33}$, $s_{12} = s_{13} = s_{23}$, and $s_{44} = s_{55} = s_{66}$; and all the remaining

TABLE 4.6 Equations for Rotated Elastic Compliances

	s'_{33}	s'_{44}	s'_{55}
s_{11}	γ_1^4	$4\gamma_1^2\beta_1^2$	*
s_{12}	$2\gamma_1^2\gamma_2^2$	$8\gamma_1\gamma_2\beta_1\beta_2$	
s_{13}	$2\gamma_1^2\gamma_3^2$	$8\gamma_1\gamma_3\beta_1\beta_3$	
s_{14}	$2\gamma_1^2\gamma_2\gamma_3$	$4\gamma_1\gamma_3\beta_1\beta_2 + 4\gamma_1\gamma_2\beta_1\beta_3$	
s_{15}	$2\gamma_1^3\gamma_3$	$4\gamma_1^2\beta_1\beta_3 + 4\gamma_1\gamma_3\beta_1^2$	
s_{16}	$2\gamma_1^3\gamma_2$	$4\gamma_1^2\beta_1\beta_2 + 4\gamma_1\gamma_2\beta_1^2$	
s_{22}	γ_2^4	$4\gamma_2^2\beta_2^2$	
s_{23}	$2\gamma_2^2\gamma_3^2$	$8\gamma_2\gamma_3\beta_2\beta_3$	
s_{24}	$2\gamma_2^3\gamma_3$	$4\gamma_2^2\beta_2\beta_3 + 4\gamma_2\gamma_3\beta_2^2$	
s_{25}	$2\gamma_1\gamma_2^2\gamma_3$	$4\gamma_1\gamma_2\beta_2\beta_3 + 4\gamma_2\gamma_3\beta_1\beta_2$	
s_{26}	$2\gamma_1\gamma_2^3$	$4\gamma_1\gamma_2\beta_2^2 + 4\gamma_2^2\beta_1\beta_2$	
s_{33}	γ_3^4	$4\gamma_3^2\beta_3^2$	
s_{34}	$2\gamma_2\gamma_3^3$	$4\gamma_2\gamma_3\beta_3^2 + 4\gamma_3^2\beta_2\beta_3$	
s_{35}	$2\gamma_1\gamma_3^3$	$4\gamma_1\gamma_3\beta_3^2 + 4\gamma_3^2\beta_1\beta_3$	
s_{36}	$2\gamma_1\gamma_2\gamma_3^2$	$4\gamma_1\gamma_3\beta_2\beta_3 + 4\gamma_2\gamma_3\beta_1\beta_3$	
s_{44}	$\gamma_2^2\gamma_3^2$	$\gamma_2^2\beta_3^2 + \gamma_3^2\beta_2^2 + 2\gamma_2\gamma_3\beta_2\beta_3$	
s_{45}	$2\gamma_1\gamma_2\gamma_3^2$	$2\gamma_1\gamma_2\beta_3^2 + 2\gamma_3^2\beta_1\beta_2 + 2\gamma_1\gamma_3\beta_2\beta_3 + 2\gamma_2\gamma_3\beta_1\beta_3$	
s_{46}	$2\gamma_1\gamma_2^2\gamma_3$	$2\gamma_1\gamma_3\beta_2^2 + 2\gamma_2^2\beta_1\beta_3 + 2\gamma_1\gamma_2\beta_2\beta_3 + 2\gamma_2\gamma_3\beta_1\beta_2$	
s_{55}	$\gamma_1^2\gamma_3^2$	$\gamma_1^2\beta_3^2 + \gamma_3^2\beta_1^2 + 2\gamma_1\gamma_3\beta_1\beta_3$	
s_{56}	$2\gamma_1^2\gamma_2\gamma_3$	$2\gamma_1^2\beta_2\beta_3 + 2\gamma_2\gamma_3\beta_1^2 + 2\gamma_1\gamma_2\beta_1\beta_3 + 2\gamma_1\gamma_3\beta_1\beta_2$	
s_{66}	$\gamma_1^2\gamma_2^2$	$\gamma_1^2\beta_2^2 + \gamma_2^2\beta_1^2 + 2\gamma_1\gamma_2\beta_1\beta_2$	

*Substitute α for β in the equation for S'_{44}.

s_{ij} are zero. Consequently, we can write for the free Young's modulus, substituting term for term in Eq. (4.33),

$$\frac{1}{E_f} = s'_{33} = s_{11}(\gamma_1^4 + \gamma_2^4 + \gamma_3^4) + 2s_{12}(\gamma_1^2\gamma_2^2 + \gamma_1^2\gamma_3^2 + \gamma_2^2\gamma_3^2)$$
$$+ s_{44}(\gamma_1^2\gamma_2^2 + \gamma_1^2\gamma_3^2 + \gamma_2^2\gamma_3^2)$$

which with the aid of Eqs. (4.32) and some algebra reduces to

$$\frac{1}{E_f} = s'_{33} = s_{11} + (2s_{12} - 2s_{11} + s_{44})(\gamma_1^2\gamma_2^2 + \gamma_1^2\gamma_3^2 + \gamma_2^2\gamma_3^2) \quad (4.34)$$

and, similarly for the free shear modulus, we have

$$\frac{1}{G_f} = \tfrac{1}{2}(s'_{44} + s'_{55}) = s'_{44}$$
$$= 4s_{11}(\gamma_1^2\beta_1^2 + \gamma_2^2\beta_2^2 + \gamma_3^2\beta_3^2) + 8s_{12}(\gamma_1\gamma_2\beta_1\beta_2 + \gamma_1\gamma_3\beta_1\beta_3 + \gamma_2\gamma_3\beta_2\beta_3)$$
$$+ s_{44}(\gamma_2^2\beta_3^2 + \gamma_3^2\beta_2^2 + 2\gamma_2\gamma_3\beta_2\beta_3 + \gamma_1^2\beta_3^2 + \gamma_3^2\beta_1^2$$
$$+ 2\gamma_1\gamma_3\beta_1\beta_3 + \gamma_1^2\beta_1^2 + \gamma_2^2\beta_2^2 + 2\gamma_1\gamma_2\beta_1\beta_2)$$

which upon application of Eqs. (4.33) and considerable algebra reduces to

$$\frac{1}{G_f} = s'_{44} = s_{44} + 4(s_{44} + 2s_{12} - 2s_{11})(\gamma_1\gamma_2\beta_1\beta_2 + \gamma_1\gamma_3\beta_1\beta_3 + \gamma_2\gamma_3\beta_2\beta_3) \quad (4.35)$$

In both Eqs. (4.34) and (4.35), we note that the term $(s_{44} + 2s_{12} - 2s_{11})$ goes to zero for isotropic materials, since $s_{44} = (2s_{11} - 2s_{12})$, so that, for isotropic materials,

$$\frac{1}{E_f} = s'_{33} = s_{11} \quad \text{and} \quad \frac{1}{G_f} = s'_{44} = s_{44}$$

i.e., there are only two independent compliances, and these are independent of orientation.

For tetragonal minerals, there are two classes whose elastic compliance matrices are the same, except that the matrix for the lower symmetry classes (4, $\bar{4}$, 4/m) contains the terms $s_{16} = -s_{26}$, which are both zero in the higher symmetry classes ($\bar{4}2m$, 4mm, 422, 4/m 2/m 2/m). For the classes 4, $\bar{4}$, and 4/m, symmetry imposes the following conditions:

$$s_{11} = s_{22}, \, s_{13} = s_{23}, \, s_{44} = s_{55}, \, s_{16} = -s_{26}$$

and $\quad s_{14} = s_{15} = s_{24} = s_{34} = s_{35} = s_{36} = s_{45} = s_{46} = s_{56} = 0$

Proceeding in the same manner as for the cubic case, we can write, by direct substitution into Eq. (4.33),

$$\frac{1}{E_f} = s'_{33} = s_{11}\gamma_1^4 + 2s_{12}\gamma_1^2\gamma_2^2 + 2s_{13}\gamma_1^2\gamma_3^2 + 2s_{16}\gamma_1^3\gamma_2 + s_{11}\gamma_2^4 + 2s_{13}\gamma_2^2\gamma_3^2$$
$$+ s_{33}\gamma_3^4 + s_{44}\gamma_2^2\gamma_3^2 + s_{44}\gamma_1^2\gamma_3^2 + s_{66}\gamma_1^2\gamma_2^2 - 2s_{16}\gamma_1\gamma_2^3 \quad (4.36)$$

which can be reduced to

$$\frac{1}{E_f} = s'_{33} = s_{11}(1 - \gamma_3^2)^2 + s_{33}\gamma_3^4 + (2s_{13} + s_{44})(\gamma_3^2 - \gamma_3^4)$$
$$+ (2s_{12} - 2s_{11} + s_{66})\gamma_2^2\gamma_1^2 + 2s_{16}\gamma_1\gamma_2(\gamma_1^2 - \gamma_2^2) \quad (4.37)$$

and, by repeating the procedure, one can obtain an expression for $1/G_f = (s'_{44} + s'_{55})/2$, analogous to Eq. (4.36), which may then be reduced to

$$\frac{1}{G_f} = \frac{s'_{44} + s'_{55}}{2} = \tfrac{1}{2}s_{66}(1 - \gamma_3^2) + \tfrac{1}{2}s_{44}(1 + \gamma_3^2)$$
$$+ 2(s_{11} + s_{33} - s_{44} - 2s_{13})(\gamma_3^2 - \gamma_3^4)$$
$$+ 2(2s_{11} - 2s_{12} - s_{66})\gamma_2^2\gamma_1^2 - s_{16}\gamma_1\gamma_2(\gamma_1^2 - \gamma_2^2) \quad (4.38)$$

Since the four classes of higher tetragonal symmetry have an identical elastic compliance matrix, except that $s_{16} = -s_{26} = 0$, the relations expressing the free Young's and shear moduli are identical to Eqs. (4.37) and (4.38), with the last terms on the right containing s_{16}, zero.

The reader may now note that the equations for cubic materials (4.34) and (4.35) could be obtained by the reduction of Eqs. (4.37) and (4.38) by following the reductions in the elastic-compliance matrices which is imposed by the increased symmetry of the cubic system.

Equations (4.34) and (4.35) involve three independent elastic compliances. Consequently, the data of Young's and shear moduli measured on two specimens oriented in two different directions are needed to determine the three elastic compliances in the cubic system. The number of specimens required to determine all the elastic compliances increases with increasing complexity of the symmetry system. For example, with minerals in the tetragonal system, such as rutile, at least four specimens are required. A discussion of the solutions for the tetragonal and hexagonal systems is given by Wachtman et al.[22-24]

4.10 Measurements at High Temperatures

One of the advantages of the standard resonance method is that the system is suitable for measuring the elastic moduli of solids as a function of temperature. This is due to the fact that the driving and pickup transducers can easily be located outside the furnace, so that only the specimen is exposed to high temperatures, and also that both flexural and torsional resonant frequencies can be determined together. The application of this system to high-temperature work is due to Förster,[25,26] who measured the change in Young's modulus and internal friction of metals up to 1000°C. Since little special care is needed other than for a good temperature distribution within the oven, this system is widely used. A series of work undertaken by Wachtman, Spinner, and their associates[27-32] at the National Bureau of Standards was performed employing a similar system. Work by Chung and Lawrence[33] and by Soga and Anderson[34,35] was also done with similar equipment.

The following care must be taken in the meausrement. (1) The temperature of the specimen should be as uniform as possible from one end to the other, preferably within 2°C. (2) The support wires, if they are of refractory metals such as nichrome, kanthal, or tungsten, might exhibit their own resonant condition near the resonant frequency of the specimen. Therefore, it is advisable to use glass threads instead of metal wires to as high a temperature as can be reached (possibly 700 to 800°C, using fused-silica threads) in order to cross-check the results.

The elastic moduli at high temperatures can be calculated from a knowledge of the temperature dependence of the resonant frequencies and the thermal expansion of the solid. In the case of isotropic and cubic solids, the density at high temperature is expressible by

$$\rho^T = \rho^{298}\left(\frac{l^{298}}{l^T}\right)^3 \tag{4.39}$$

where the superscripts indicate values at 298°K and T°K.

Since the shear modulus can be expressed by Eq. (4.3) and the shape of the cross section of the specimen remains the same, the relative change in the shear modulus ($F_G{}^T$) with temperature becomes

$$F_G{}^T = \frac{G^T}{G^{298}} = \left(\frac{f_G{}^T}{f_G{}^{298}}\right)^2 \frac{l^{298}}{l^T} \tag{4.40}$$

where f_G is the torsional resonant frequency. We emphasize that this relationship holds for isotropic and cubic solids only.

In the case of Young's modulus, the situation is not as simple, since the equation relating the flexural vibration and Young's modulus involves the shape factor T, which is a function of Poisson's ratio as well as the shape of the specimen (as shown in Sec. 4.4). The relative change in Young's modulus ($F_E{}^T$) with temperature is, therefore,

$$F_E{}^T = \frac{E^T}{E^{298}} = \left(\frac{f_E{}^T}{f_E{}^{298}}\right)^2 \left(\frac{l^{298}}{l^T}\right)\left(\frac{T^T}{T^{298}}\right) \tag{4.41}$$

In the special case where Poisson's ratio remains constant—the condition for this is $(f_E{}^T/f_E{}^{298}) = (f_G{}^T/f_G{}^{298})$—the last term can be dropped. In actual practice, the change in T^T/T^{298} may be neglected except for solids with a large change in Poisson's ratio. For such a case, one first calculates the change in Poisson's ratio from the frequency ratio

$$\frac{(\sigma^T - 1)}{(\sigma^{298} - 1)} = \left(\frac{f_E{}^T}{f_E{}^{298}}\right)^2 \left(\frac{f_G{}^{298}}{f_G{}^T}\right)^2 \tag{4.42}$$

Then $F_E{}^T$ is calculated by using an appropriate T^T for σ^T instead of T^{298}. Generally, however, the difference in the values of F^T based on T^T and T^{298} is very small (less than 0.1 percent).

For an isotropic solid, the adiabatic bulk modulus can be calculated from Young's modulus and the shear modulus with the equation

$$B = \frac{EG}{3(3G - E)} \tag{4.43}$$

From Eqs. (4.40) to (4.42), the relative change in the bulk modulus with temperature can be expressed in the form

$$F_B^T = \frac{B^T}{B^{298}} = F_E^T \left\{ 1 - \left[\frac{(F_E^T/F_G^T) - 1}{(3G^{298}/E^{298}) - 1} \right] \right\}^{-1} \quad (4.44)$$

By appropriately suspending the specimen, the flexural and torsional resonant frequencies can be determined simultaneously at one temperature,

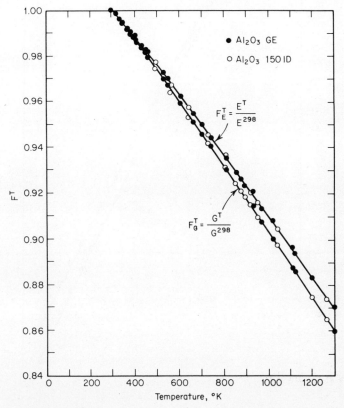

Fig. 4.4 The relative change in Young's modulus and shear modulus with temperature for polycrystalline Al_2O_3.

so Eq. (4.44) makes it possible to compute the relative change in adiabatic bulk modulus with temperature directly from the relative changes in Young's modulus and the shear modulus. Examples of the actual measurements for polycrystalline MgO and Al_2O_3 are shown in Figs. 4.4 to 4.6.[34] As expected from Eq. (4.44) slight variations in F_E and F_G result in large variation in F_B.

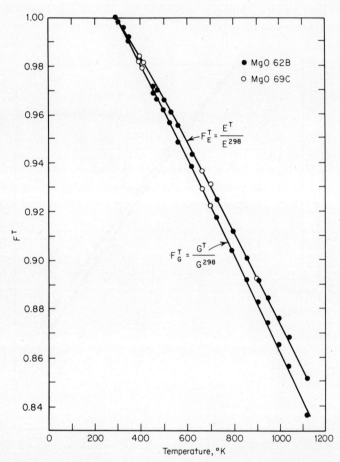

Fig. 4.5 Relative change in Young's modulus and shear modulus with temperature for polycrystalline MgO.

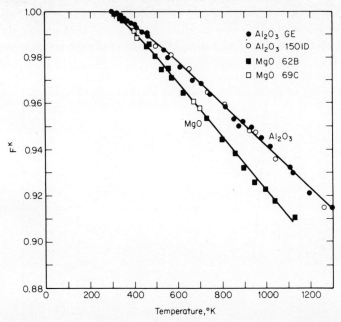

Fig. 4.6 Relative change in bulk modulus with temperature for polycrystalline MgO and Al_2O_3.

4.11 Vibration of a Wire or Bar Clamped at One End

From the forced resonance of transverse oscillation of a vibrating cantilever, we can determine the dynamic value of Young's modulus. The theory of this vibrating cantilever method has been worked out in detail by Prescott.[36] The frequency of the fundamental normal mode f_0 is given by

$$\frac{K}{2\pi}\left(\frac{E}{\rho}\right)^{1/2} = \frac{l^2 f_0}{Z^2} \qquad (4.45)$$

where l = length of the rod
ρ = density
E = Young's modulus
K = radius of gyration of the cross section perpendicular to its plane of motion
Z = a parameter which is a function of the ratio C of the mass of the excitation device attached on the rod to the mass of the rod, and is given by

$$\frac{1 + \cosh Z \cos Z}{\cosh Z \sin Z - \sinh Z \cos Z} = CZ \qquad (4.46)$$

The solution of this equation can be found by plotting the values of the left-hand side of the equation as a function of Z and then finding the values of Z which intersect with the straight line $y = CZ$; the result is presented in Fig. 4.7.

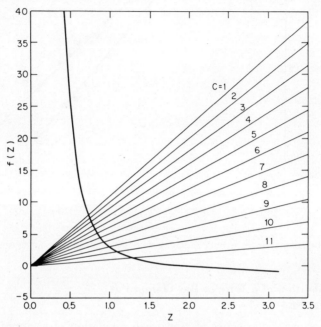

Fig. 4.7 Graph of $f(z)$ vs. Z for use in application of Eq. (4.46).

In order to excite the specimen and detect its resonant condition, a small magnetic pole piece may be attached on the one free end of the rod and set in forced vibration by a magnetic coil placed nearby. This method was successfully used by Davies and James,[37] Hillier,[38,39] and others. If the displacement of the bar is ξ and a driving force F is imposed on it, the equation of motion is given by

$$m\frac{d^2\xi}{dt^2} + q\frac{d\xi}{dx} + s\xi = F \tag{4.47}$$

where s = elastic compliance
m = mass
q = resistance coefficient

At the resonant condition with a frequency f_0, the force F is zero for a material of low acoustic loss. The amplitude of the displacement ξ is observed as a function of frequency. Since the maximum value of ξ_m

occurs at a frequency f_m slightly less than f_0, some extrapolation is needed to obtain the value of f_0. This can be done by obtaining values for f_m at different field strengths and extrapolating it to zero field strength. As a consequence of this procedure, the accuracy of the data is about 0.2 percent.[38]

The shear modulus of a material can be obtained by finding the periodic time of torsional oscillations with a very thin wire or rod. To maintain the oscillations for a considerable period of time, a heavy weight is attached to one end of the wire. When the weight is twisted through an angle Φ, the wire will have a torque applied of $(G\Phi/l)(\pi r^4/2)$, where l and r are the length and radius of the wire, respectively. The motion of the weight is a simple harmonic motion given by

$$I\frac{d^2\Phi}{dt^2}+\frac{G}{l}\frac{\pi r^4}{2}\Phi=0 \qquad (4.48)$$

where I is the moment of inertia of the weight. The period T is, therefore,

$$T=2\pi\left(\frac{2Il}{G\pi r^4}\right)^{1/2} \qquad (4.49)$$

When the moment of inertia of the weight is unknown because of its irregular shape, then an additional body of known moment of inertia I_1 is attached to the system. The period T_1 is then given by

$$T_1=2\pi\left[\frac{2(I_1+I)l}{G\pi r^4}\right]^{1/2} \qquad (4.50)$$

Substituting I with other parameters, we get

$$T_1^2-T^2=\frac{8\pi I_1 l}{Gr^4} \qquad (4.51)$$

The radius appears in the equation raised to the fourth power; consequently, the value of G depends largely on the uniformity of the wire, or the accuracy of the radius measurement. The accurate determination of the absolute shear modulus is therefore rather difficult. However, the data are suitable for determining the relative change in the shear modulus with temperature,[40] and Eq. (4.49) can be used to obtain the shear modulus at different temperatures. Because of the low level of stress applied, this torsional method is also quite useful for obtaining internal friction at very low frequencies. Figure 4.8 shows a typical experimental setup used for such a purpose.[41] The initial twist of the inertial bar can be given by using a pair of electromagnets. A mirror attached to the bar reflects a light beam, and this can be recorded photographically onto a chart moving with a known speed to obtain the period of vibration. It should be

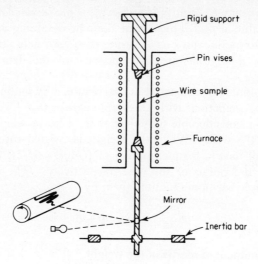

Fig. 4.8 Torsion pendulum.

pointed out that in the torsional method, strain in the specimen is not homogeneous. In actual experiments, if the applied stress is chosen within a range such that the internal friction and the dynamic shear modulus are independent of the amplitude of vibration,[41] the effect of stress inhomogeneity may be ignored.

4.12 Other Methods of Employing Resonance

The application of the standard resonance method for determining elastic properties of materials is limited by the size of the specimen measured. When the length becomes shorter than 3 in, the torsional fundamental resonance frequency of materials with large moduli (such as Al_2O_3, MgO, or SiC) exceeds 40 kHz, so that the usual equipment is not adequate. Moreover, all the dimensions (length, width, and thickness) are critically involved in the calculation of the elastic moduli from the resonance frequency, as shown in Secs. 4.3 and 4.4. From a practical point of view, it is quite difficult to fabricate small rods or bars with a uniform cross-sectional dimension. However, one sometimes encounters the situation where the elastic moduli must be determined from small specimens.

In order to deal with this problem, several methods of determining the elastic moduli of solids using rods or bars in resonance may be used. Most of these methods were developed to determine the internal friction of the solids rather than measuring elastic moduli. Depending upon the choice of excitation and detection of the resonant condition, they may be classified as (1) piezoelectric effect, (2) electromagnetic effect, and (3) electrostatic effect. In the case of nonmetallic compounds, it is rather difficult to set a specimen in resonance by these techniques without applying a conductive film or attaching transducers. Consequently, a correction must be made in order to obtain the absolute value of elastic moduli of the specimen. Because of the difficulty in making such corrections, we should not expect to determine the elastic moduli accurately unless we know all the effects. In the following section, we describe some of the typical methods in each category.

4.12.1 Piezoelectric Effect. A piezoelectric transducer is cemented to the rod or bar and this composite system is made to resonate. This method seems to have started from the work by Quimby,[42] who used a longitudinal-mode quartz transducer cemented to the specimen with a very thin layer of shellac. This work was followed by Balamuth,[43] Rose,[44] Norwich,[45] and Sutton.[46]

When the specimen, in the form of a thin cylinder, is cemented to a cylindrical transducer of equal cross section, the mechanical resonant frequencies of the composite cylinder are given by[43,47]

$$v_1 \rho_1 \tan\left(\frac{2\pi f l_1}{v_1}\right) + v_2 \rho_2 \tan\left(\frac{2\pi f l_2}{v_2}\right) = 0 \qquad (4.52)$$

where ρ_1 and ρ_2, l_1 and l_2 = densities and lengths of the specimen and transducer, respectively

v_1 = longitudinal or torsional velocity in the direction of the cylinder axis

v_2 = longitudinal or torsional velocity of transducer

f = longitudinal or torsional resonant frequency of the composite cylinder

Since $v_i = 2 l_i f_i$, the above equation can be expressed as

$$M_1 f_1 \tan\left(\frac{2\pi f}{f_1}\right) + M_2 f_2 \tan\left(\frac{2\pi f}{f_2}\right) = 0 \qquad (4.53)$$

where M_1 and M_2 are the mass of the specimen and transducer. The resonant frequency f_1 of the specimen, therefore, can be calculated from M_1, M_2, f_2, and f. It may be possible to obtain the absolute elastic moduli with an accuracy of about 0.1 percent.[47]

The general arrangement for observing the resonance of the composite oscillator is shown in Fig. 4.9. When the frequency of the applied voltage approaches a frequency of free vibration of the composite oscillator, the

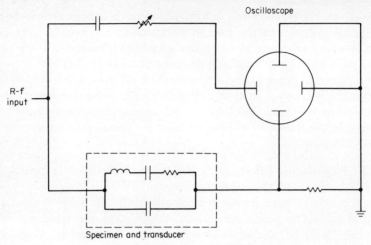

Fig. 4.9 Simple circuit for the composite oscillator.

electrical characteristic of the oscillator may be regarded as a fixed capacity shunted by an inductance, capacitance, and resistance in series. Since the actual frequency of free vibration is identical with the resonant frequency of the series branch alone, it can be deduced from the observed electric behavior of the oscillator. The output current will be a maximum at the resonant frequency of the composite f.

The torsional vibration can be achieved by placing a Y-cut quartz transducer at the end. The actual types of transducers used are shown in Fig. 4.10. This system can be operated at very high frequencies.

For the accurate determination of elastic moduli, it is necessary to consider the effect of the cement applied between the specimen and the transducer. Terry[47] regarded the cement as a third cylinder and estimated the

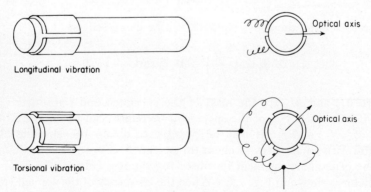

Fig. 4.10 Quartz transducer for the composite oscillator.

error in the calculated value of the elastic wave velocity v_1 due to neglecting the cement. The error Δv is given by

$$\Delta v = \frac{M_3}{M_1}\left[\cos^2\theta_1 + \left(\frac{\rho_1 v_1}{\rho_3 v_3}\right)^2 \sin^2\theta_1\right]\left(\frac{2\theta_1}{2\theta_1 - \sin 2\theta_1}\right) \quad (4.54)$$

where $\theta_i = 2\pi f/f_i$ and the subscripts 1 and 3 refer to the specimen and the cement, respectively. If the composite rod is vibrated in its fundamental mode and the specimen and transducer have been selected to have approximately equal natural frequencies, so that $f_1 \simeq f_2 \simeq 2f$, the error due to the cement can be expressed as

$$\Delta v \simeq -\frac{M_3}{M_1}\left(\frac{\rho_1 v_1}{\rho_3 v_3}\right)^2 \quad (4.55)$$

Since $\rho v^2 = C$ and $M/\rho = \pi r^2 l$ for a cylinder, Eq. (4.55) becomes

$$\Delta v \simeq -\frac{l_3}{l_1}\frac{C_1}{C_3} \quad (4.56)$$

where C is the appropriate elastic modulus.

Marx[48] used the three-component resonator, which consists of a specimen, driver, and gauge. The addition of an auxiliary quartz crystal as the gauge enabled him to determine instantaneous values of the internal friction of glass with rising temperature up to the softening point of the glass.[49]

4.12.2 Electromagnetic Effect. The simplest system which utilizes the magnetic effect to vibrate the specimen is that of Wegel and Walther.[50] Two thin steel magnetic armatures were cemented to both ends of the specimen, and the composite was placed in an alternating magnetic field. Since the mass of this armature must be taken into account, several improvements have been made to avoid the use of the armature. The basic system shown in Fig. 4.11a, which is based on the effect of eddy currents, was given by Zener et al.,[51] who studied the elastic and anelastic properties of metals   A permanent magnet applies a force on the alternating currents passing through the coil surrounding the specimen. The detector operates on the eddy currents induced at the end of the bar due to longitudinal (or extensional) motion in the magnetic field. Fusfeld[52] describes a system capable of rapid determination of internal friction based upon the same principle. The article by Hanlon and Wolf[53] illustrates how to measure the torsional vibration, as shown in Fig. 4.11b.

Bradfield[54] gives a method based on the inductor principle. The end of a specimen in the form of a thin strip is placed between two coils which are set very closely in a strong magnetic field. A powerful field induces eddy currents within the strip, resulting in a longitudinal resonance vibration of the strip. The resonance condition is detected with a bridge

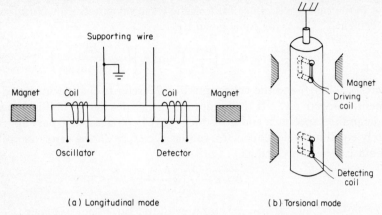

Fig. 4.11 System for driving a conductive specimen with eddy current.

circuit, and Bradfield was successful in measuring the resonant frequencies of very small specimens ($20 \times 3 \times 0.05$ mm). Bradfield[55] also reviewed the method of using magnetostrictive transducers to measure elasticity dynamically. The excitation and detection of the system are the same as described in Sec. 4.6, but may be usable up to 1 MHz, depending on the transducer.

Saint Clair[56] used a method where the vibration is caused by an electrodynamic effect linked with the setting up of induced currents generated by a variable magnetic field. The method of Barone and Giacomini[57] does not involve a variable magnetic field. Instead, they used an alternating current to set the specimen in vibration, as shown in Fig. 4.12. The surface of the specimen is partially coated with the conductive surface as shown in the figure. The vibration is caused by electrodynamic forces which are proportional to the magnetic induction; the current intensity and its frequency are the same as the alternating current. One advantage of this

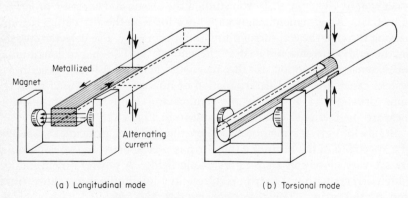

Fig. 4.12 The electrodynamic drive.

method is the large excitation force developed, so that this method is applicable to highly damped solids.

4.12.3 Electrostatic Effect. When the oscillator voltage exciting the sample is applied between the conductive surface of the specimen and a closely adjacent electrode, the specimen vibrates through its electrostatic attraction. This method originated with the work of Bancroft and Jacobs.[58] An arrangement to excite three different vibrations is shown in Fig. 4.13.

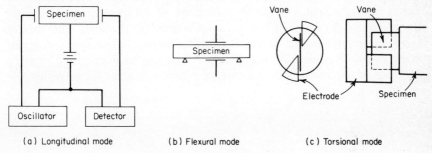

Fig. 4.13 The electrostatic drive.

A vane is needed to excite torsional vibrations, as shown by Vernon.[59] This affects the resonant frequency of the specimen as given by

$$f = \frac{1}{2l}\left(1 - \frac{I_v}{I_s}\right)\left(\frac{G}{\rho}\right)^{1/2} \quad (4.57)$$

where I_v = moment of inertia of the vane
I_s = moment of inertia of the specimen

However, for such a small vane, I_v is sufficiently small compared with the I_s of the specimen so that it is negligible.

Several successful attempts have been made to adapt this method to a very small specimen. Bordoni and Nuovo[60] developed a method by which they measured the longitudinal vibration of a circular plate specimen of 30-mm diameter and 6.4-mm thickness, with frequencies up to 5.5 MHz. Their arrangement is shown in Fig. 4.14. The use of mica instead of

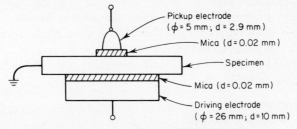

Fig. 4.14 Electrostatic drive for small specimens.

air as the dielectric medium permits the reduction of the distance between the specimen and electrode, so that higher frequencies can be employed. It also allows sources as high as 250 V to be applied from the power amplifier. Detection is performed with a frequency-modulation technique which has been improved by Vernon,[59] Pursey and Pyatt,[61] Harlow et al.,[62] and Hinton.[63] They also introduced a one-electrode system which has the advantage that only one end of the specimen need be approached, and the system becomes simple if one wants to determine resonant frequency and internal friction as functions of temperature. The one-transducer system does not need a bias potential, for the specimen vibrates at the second harmonic of the driving frequency. The block diagram of the system is shown in Fig. 4.15. According to Pursey and Pyatt,[61] the value of the true elastic moduli so obtained is accurate to within 0.2 percent.

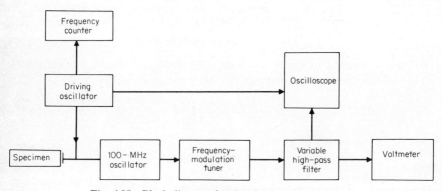

Fig. 4.15 Block diagram for the electrostatic method.

4.13 Problems Related to Internal Friction

The anelasticity or internal friction of solids is one of the important characteristics which are indirectly related to their elastic properties. Since the same apparatus described in the preceding sections can be used to determine the internal friction, a brief description is given here. To obtain a fuller knowledge of the problems, one should consult the classical book, "Elasticity and Anelasticity of Metals," by Zener.[64]

Generally, the method of measuring internal friction of solids is divided into the following:[64,65]

1. Free-vibration method
2. Forced-vibration method
3. Wave-propagation method
4. Direct observation of stress-strain curves

The free-vibration method is based on the measurement of the decay in amplitude of vibrations during free vibration. The logarithmic decrement is defined as the natural logarithm of the ratio of successive amplitudes,

$$\delta = \frac{\ln(A_1/A_2)}{(t_1 - t_2)f} \simeq \frac{1}{2}\frac{\Delta E}{E} \qquad (4.58)$$

where A_1 and A_2 = amplitude at times t_1 and t_2
f = natural frequency of the free-vibrating system
ΔE = energy lost per cycle
E = stored vibration energy

When the damping is small, the internal friction Q^{-1} is expressed by

$$Q^{-1} = \frac{\delta}{\pi} \qquad (4.59)$$

The torsional method described in Sec. 4.11 and the resonant-sphere technique in Chap. 5 belong to this group. Also, an apparatus developed by Burch et al.[66] used a composite reed-type pendulum with a small ceramic specimen of about 2.5 cm long. A linear variable differential transformer and a recorder were employed to detect the decay in amplitude of vibrations which were initiated by pulsing a small electromagnet. The apparatus was capable of determining values of the order 10^{-4}, which is comparable to the torsional method. Wachtman and Tefft[67] have shown that values of Q^{-1} as low as 5×10^{-6} may be determined by the standard resonance method if the dissipation of energy, such as acoustical radiation, to the surroundings and loss due to the restriction of free motion caused by suspension wires are eliminated. The effects of air damping and suspension on the measured values of Q were obtained as a function of suspension position. To detect the decay in amplitude, the experimenters used a storage oscilloscope, which holds the trace of a single damped sine wave indefinitely. Alternately, one may use a device to directly count the number of cycles occurring during the decrease in amplitude from 90 to 65.5 percent.

The standard resonance method can be used as a forced vibration method. The specimen is forced to oscillate at various frequencies with constant input. When the impressed frequency is equal to a critical resonance frequency f of the specimen, the amplitude of vibration becomes a maximum. This amplitude decreases as the impressed frequency deviates to either side of f. Once the change in impressed frequency necessary to alter the amplitude of vibration from half the maximum value on one side to half the maximum value on the other side of the resonant frequency is determined, the internal friction is obtained by

$$Q^{-1} = 0.5773 \frac{\Delta f}{f} \qquad (4.60)$$

This method can be applied to a specimen having large acoustic loss.

Since there are many materials having much lower values of Q^{-1} than 10^{-5}, it is necessary to apply other techniques on such materials. For this purpose, the wave-propagation method may be suitable. With this method, it was possible to investigate the effect of impurity content and of x-ray radiation on the elastic properties of synthetic quartz at low temperature.[68] One of the disadvantages of this technique has been that high-temperature effects cannot be investigated because of the limitation of transducers. However, this is being overcome with the development of high-temperature transducers. Lithium niobate,[69] which can withstand temperatures above 900°C, may provide the means to extend the measurements discussed into the realm of high temperatures.

REFERENCES

1. Ide, J. M.: Some Dynamic Methods of Determination of Young's Modulus, *Rev. Sci. Instr.*, **6**:296 (1935).
2. Spinner, S., and W. E. Tefft: Method for Determining Mechanical Resonance Frequencies and for Calculating Elastic Moduli from These Frequencies, *Am. Soc. Test. Mater. Proc.*, **61**:1221 (1961).
3. Pochhammer, L.: Über die Fortpflanzungsgeschwindigkeit kleiner Schwingungen in einem unbegrentzen Isotropen Kreiscylinder, *J. Reine u. Angew Math.*, **81**:324 (1876).
4. Davies, R. M.: The Frequency of Longitudinal and Torsional Vibration of Unloaded and Loaded Bars, *Philos. Mag.*, **25**:364 (1938).
5. Tefft, W. E., and S. Spinner: Torsional Resonance Vibrations of Uniform Bars of Square Cross Section, *J. Res. Natl. Bur. Stand.*, **65A**:167 (1961).
6. Timoshenko, S., and J. N. Goodier: "Theory of Elasticity," 2d ed., p. 277, McGraw-Hill, New York, 1951.
7. Pickett, G.: Equations for Computing Elastic Constants from Flexural and Torsional Resonant Frequencies of Vibration of Prisms and Cylinders, *Am. Soc. Test. Mater. Proc.*, **45**:846 (1945).
8. Hasselman, D. P. H.: "Tables for the Computation of the Shear Modulus and Young's Modulus of Elasticity from the Resonant Frequencies of Rectangular Prisms," Carborundum Co., Niagara Falls, N.Y., 1961.
9. Spinner, S., and R. C. Valore, Jr.: Comparison of Theoretical and Empirical Relations between the Shear Modulus and Torsional Resonance Frequencies for Bars of Rectangular Cross Section, *J. Res. Natl. Bur. Stand.*, **60A**:459 (1958).
10. Goens, E.: Uber die Bestimmung des Elastizitätsmodulus von Stäben mit Hilfe von Biegungschwingen, *Ann. Phys.*, B. Folge, Band **11**:649 (1931).
11. Tefft, W. E.: Numerical Solution of the Frequency Equations for the Flexural Vibration of Cylindrical Rods, *J. Res. Natl. Bur. Stand.*, **64B**:237 (1960).
12. Spinner, S., T. W. Reichard, and W. E. Tefft: A Comparison of Experimental and Theoretical Relations between Young's Modulus and Flexural and Longitudinal Resonance Frequencies of Uniform Bars, *J. Res. Natl. Bur. Stand.*, **64A**:147 (1960).
13. Rayleigh, Lord: "The Theory of Sound," vol. 1, Dover, New York, 1945.
14. Love, A. E. H.: "A Treatise on the Mathematical Theory of Elasticity," 4th ed., Dover, New York, 1944.
15. Lamb, H.: "The Dynamical Theory of Sound," 2d ed., Dover, New York, 1960.

16. Bancroft, D.: The Velocity of Longitudinal Waves in Cylindrical Bars, *Phys. Rev.*, **59**:588 (1941).
17. Voigt, W.: "Lehrbuch der Kristallphysik," Teubner, Leipzig, 1928.
18. Goens, E.: Torsional and Flexural Oscillations of a Monocrystalline Rod, *Ann. Phys.*, **15**:455 (1932).
19. Brown, Jr., W. F.: Interpretation of Torsional Frequencies of Crystal Specimens, *Phys. Rev.*, **58**:998 (1940).
20. Hearmon, R. F. S.: The Elastic Constants of Anisotropic Materials, *Rev. Mod. Phys.*, **18**:409 (1946).
21. Hearmon, R. F. S.: The Elastic Constants of Anisotropic Materials II, *Adv. Phys.*, **5**:323 (1956).
22. Wachtman, Jr., J. B., W. E. Tefft, and D. G. Lam, Jr.: Elastic Constants of Rutile (TiO_2), *J. Res. Natl. Bur. Stand.*, **66A**:465 (1962).
23. Wachtman, Jr., J. B., W. S. Brower, Jr., and E. N. Farabaugh: Elastic Constants of Single Crystal Calcium Molybdate, ($CaMoO_4$), *J. Am. Ceram. Soc.*, **51**:341 (1968).
24. Wachtman, Jr., J. B., W. E. Tefft, D. G. Lam, Jr., and R. P. Stinchfield: Elastic Constants of Synthetic Single Crystal Corundum at Room Temperature, *J. Res. Natl. Bur. Stand.*, **64A**:213 (1960).
25. Förster, F.: Ein neues Messverfahren zur Bestimmung des Elastizitätsmodulus und der Dämpfung, *Z. Metallkd.*, **29**:109 (1937).
26. Förster, F., and W. Köster: Elasticity Modulus and Damping Capacity Related to the Condition of the Material, *Z. Metallkd.*, **29**:116 (1937).
27. Wachtman, Jr., J. B., and D. G. Lam, Jr.: Young's Modulus of Various Refractory Materials as a Function of Temperature, *J. Am. Ceram. Soc.*, **42**:254 (1959).
28. Wachtman, Jr., J. B., W. E. Tefft, D. G. Lam, Jr., and C. S. Apstein: Exponential Temperature Dependence of Young's Modulus for Several Oxides, *Phys. Rev.*, **122**:1754 (1961).
29. Spinner, S.: Temperature Dependence of Elastic Constants of Some Cermet Specimens, *J. Res. Natl. Bur. Stand.*, **65C**:89 (1961).
30. Spinner, S.: Temperature Dependence of Elastic Constants of Vitreous Silica, *J. Am. Ceram. Soc.*, **45**:394 (1962).
31. Spinner, S., L. Stone, and F. P. Knudsen: Temperature Dependence of the Elastic Constants of Thoria Specimens of Varying Porosity, *J. Res. Natl. Bur. Stand.*, **67C**:93 (1963).
32. Wachtman, Jr., J. B., and S. Spinner: Some Elastic Compliances of Single Crystal Rutile from 25 to 1000°C, *J. Res. Natl. Bur. Stand.*, **68A**:669 (1964).
33. Chung, D. H., and W. G. Lawrence: Relation of Single-Crystal Elastic Constants to Polycrystalline Isotropic Elastic Moduli of MgO: II. Temperature Dependence, *J. Am. Ceram. Soc.*, **47**:448 (1964).
34. Soga, N., and O. L. Anderson: High-Temperature Elastic Properties of Polycrystalline MgO and Al_2O_3, *J. Am. Ceram. Soc.*, **49**:355 (1966).
35. Soga, N., and O. L. Anderson: High-Temperature Elasticity and Expansivity of Forsterite and Steatite, *J. Am. Ceram. Soc.*, **50**:239 (1967).
36. Prescott, "Applied Elasticity," Longmans, London, 1924.
37. Davies, R. M., and E. G. James: A Study of an Electrically-Maintained Vibrating Reed and Its Application to the Determination of Young's Modulus, *Philos. Mag.*, **18**:1023 (1934).
38. Hillier, K. W.: A Vibrating Cantilever Method for Investigation of the Dynamic Elasticity of High Polymers, *Proc. Phys. Soc.*, **64B**:998 (1951).
39. Hillier, K. W.: A Review of the Progress in the Measurement of Dynamic Elastic Properties, in N. Davids (ed.), "International Symposium on Stress Wave Propagation in Materials," Interscience, New York, 1960.

40. Kingery, W. D.: "Property Measurements at High Temperature," Wiley, New York, 1959.
41. Kê, T. S.: Experimental Evidence of the Viscous Behavior of Grain Boundaries in Metals, *Phys. Rev.*, **71**:533 (1947).
42. Quimby, S. L.: On the Experimental Determination of the Viscosity of Vibrating Solids, *Phys. Rev.*, **25**:558 (1925).
43. Balamuth, L.: A New Method for Measuring Elastic Moduli and the Variation with Temperature of the Principal Young's Modulus of Rocksalt between 78°K and 273°K, *Phys. Rev.*, **45**:715 (1934).
44. Rose, F. C.: The Variation of the Adiabatic Elastic Moduli of Rocksalt with Temperature between 80°K and 270°K, *Phys. Rev.*, **49**:50 (1936).
45. Norwich, A. S.: Variation of Amplitude-Dependent Internal Friction in Single Crystals of Copper with Frequency and Temperature, *Phys. Rev.*, **80**:249 (1950).
46. Sutton, P. M.: The Variation of the Elastic Constants of Crystalline Aluminum with Temperature between 63°K and 773°K, *Phys. Rev.*, **91**:816 (1953).
47. Terry, N. B.: Some Considerations of the Magnetostrictive Composite Oscillator Method for the Measurement of Elastic Moduli, *Br. J. Appl. Phys.*, **8**:270 (1957).
48. Marx, J. W.: Use of the Piezoelectric Gauge for Internal Friction Measurements, *Rev. Sci. Instr.*, **22**:503 (1951).
49. Marx, J. W., and J. M. Sivertsen: Temperature Dependence of the Elastic Moduli and Internal Friction of Silica and Glass, *J. Appl. Phys.*, **24**:81 (1953).
50. Wegel, R. L., and H. Walther: Internal Dissipation in Solids for Small Cyclic Strains, *Physics*, **6**:141 (1935).
51. Zener, C., F. C. Rose, and R. H. Randal: Intercrystalline Thermal Currents as a Source of Internal Friction, *Phys. Rev.*, **56**:343 (1939).
52. Fusfeld, H. I.: Apparatus for Rapid Measurement of Internal Friction, *Rev. Sci. Instr.*, **21**:612 (1950).
53. Hanlon, J. E., and J. D. Wolf: Internal Friction Apparatus Using Electromagnetic Drive and Pick-up, *Rev. Sci. Instr.*, **37**:676 (1966).
54. Bradfield, G.: Ultrasonic Electro-Acoustics, *Acoustica*, **4**:171 (1954).
55. Bradfield, G.: Dynamic Measurement of Elasticity Using Resonance Method, *J. Appl. Phys.*, **11**:478 (1960).
56. Saint Clair, H. W.: An Electromagnetic Sound Generator for Producing Intense High Frequency Sound, *Rev. Sci. Instr.*, **12**:250 (1941).
57. Barone, A., and A. Giacomini: Experiments on Some Electrodynamic Ultrasonic Vibrators, *Acoustica*, **4**:182 (1954).
58. Bancroft, D., and R. B. Jacobs, An Electrostatic Method of Measuring Elastic Constants, *Rev. Sci. Instr.*, **9**:279 (1938).
59. Vernon, E. V.: An Apparatus for the Determination of Dynamic Elastic Moduli at Low Strains, *J. Sci. Instr.*, **35**:28 (1958).
60. Bordoni, P. G., and M. Nuovo: Longitudinal Vibration Measurements in the Megacycle Range Made by Electrostatic Drive and Frequency Modulation Detection, *Acoustica*, **7**:1 (1957).
61. Pursey, H., and E. C. Pyatt: An Improved Method of Measuring Dynamic Elastic Constants, using Electrostatic Drive and Frequency-Modulation Detection, *J. Sci. Instr.*, **31**:248 (1956).
62. Harlow, R. G., T. Hilton, and J. G. Rider: A Frequency Modulation Technique for the Measurement of Internal Friction, *J. Sci. Instr.*, **39**:598 (1962).
63. Hinton, T.: Measurement of Internal Friction in the Kilocycle Frequency Range, *Rev. Sci. Instr.*, **36**:1114 (1965).
64. Zener, C.: "Elasticity and Anelasticity of Metals," The University of Chicago Press, Chicago, 1948.

65. Kolsky, H.: "Stress Waves in Solids," Chaps. 5 and 6, pp. 99–162, Oxford University Press, London, 1953.
66. Burch, J. D., R. D. Carnahan, and J. O. Brittain: Improved Internal Friction Apparatus, *Rev. Sci. Instr.*, **38**:1785 (1967).
67. Wachtman, Jr., J. B., and W. E. Tefft: Effect of Suspension Position on Apparent Values of Internal Friction Determined by Förster's Method, *Rev. Sci. Instr.*, **89**:517 (1958).
68. Mason, W. P.: "Physical Acoustics and Properties of Solids," p. 289, Van Nostrand, Princeton, N.J., 1958.
69. Fraser, D. B., and A. W. Warner: Lithium Niobate: A High Temperature Piezoelectric Transducer Material, *J. Appl. Phys.*, **37**:3853 (1966).

CHAPTER FIVE

Resonant-Sphere Methods for Measuring the Velocity of Sound

5.1 Introduction

This chapter will be concerned with the problem of detecting the sound velocities of solids of spherical shape using the resonant-sphere technique. Because of the applicability to specimens of small size, emphasis will be upon the measuring system, suitable for a sphere with a diameter of a few millimeters.

The resonant-sphere technique, developed by Fraser and LeCraw,[1] is a new addition to resonance methods for determining elastic constants of solids from their mechanical resonance frequencies. A major difference between this method and the standard dynamic resonance method, discussed in Chap. 4, is the shape of specimen used: a sphere versus a cylinder or rectangular prism. Since one can make a very small sphere accurately by using one of the methods described later, the accuracy of the elastic-constant data is very precise even though a small specimen is used. Errors in the measurements arise mainly from uncertainty concerning the diameter, not from the electronic measuring system, or from the equations used.

In the case of a sphere, the principal modes of vibration are torsional (or toroidal) and spheroidal (or poloidal). The torsional oscillations are those in which a particle executes motion on a spherical surface, and the spheroidal oscillations are those in which a particle executes both radial and tangential motion. For a sphere with a diameter of a few millimeters, the resonant frequencies of the fundamental modes of these oscillations lie in the range of frequencies from 0.5 to 3 MHz.

When a sphere is shaped from a single crystal, numerous modes appear. The analysis of the free oscillation of such a sphere is complicated because the resonant conditions depend on the crystallographic orientation of the specimen. For an isotropic solid, such as glass or a homogeneous polycrystalline material, the analysis is simplified because it follows the mathematical solutions given for the vibration of an elastic, homogeneous, isotropic sphere. The mathematical solutions for the vibration of an elastic anisotropic sphere are not currently available.

5.2 Free Oscillations of a Sphere

Solutions of the equation of motion of an isotropic and homogeneous elastic sphere have been dealt with by several authors since Kelvin[2] investigated it to solve the rigidity of the earth in 1863. The detailed derivation of the equations quoted here can be found in the excellent articles by Lamb[3] or by Love.[4]

The equations of small motions of an isotropic elastic body are generally given by[4]

$$\rho\left(\frac{\partial^2 u}{\partial t^2}, \frac{\partial^2 v}{\partial t^2}, \frac{\partial^2 w}{\partial t^2}\right) = (\lambda + G)\left(\frac{\partial \Delta}{\partial x}, \frac{\partial \Delta}{\partial y}, \frac{\partial \Delta}{\partial z}\right) + G\nabla^2(u, v, w) \quad (5.1)$$

where u, v, and $w =$ components of displacement U, in the x, y, and z directions

G and $\lambda =$ shear modulus and Lamé's constant

The definition of Δ is given by

$$\Delta = \frac{\partial u}{\partial x} + \frac{\partial v}{\partial y} + \frac{\partial w}{\partial z} \quad (5.2)$$

For an elastic body where the displacement U is periodic, the expression in terms of frequency ω is

$$U = Ae^{2\pi i \omega t} \quad (5.3)$$

and Eq. (5.1) can be written in the following simple form,

$$(\nabla^2 + h^2)\Delta = 0 \tag{5.4}$$

where
$$h = \frac{2\pi\omega^2\rho}{\lambda + 2G} \tag{5.5}$$

For a body with spherical symmetry, the solution of Eq. (5.4) can be separated into two functions: one dependent upon the radius and another dependent upon the angular coordinates. The angular function can be expressed as a sum of surface spherical harmonics by $\rho_l^m(\cos\theta)e^{im\theta}$, where $\rho_l^m(\cos\theta)$ is the associated Legendre function as defined by Ferrer, and l and m are the integers denoting the order of the spherical harmonic with respect to the angular coordinates θ and Φ. The integers l and m determine the surface pattern of particle displacement, as shown in Fig. 5.1. Besides these two coordinates, we need another integer giving the

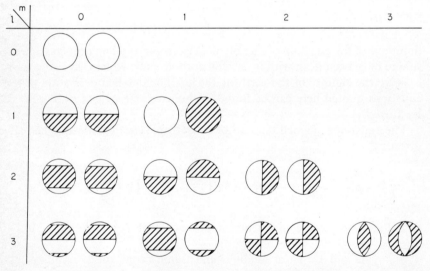

Fig. 5.1 Surface pattern of particle oscillation for a sphere.

nodal surfaces associated with the radial function. This is associated with the parameter n, which expresses the number of radial nodal surfaces. Any vibration of a sphere can be described by using these three coordinates. Among others, the notation widely used is of the form $_nS_l^m$ for the spheroidal modes and $_nT_l^m$ for the torsional modes.

In torsional vibrations, for which $\Delta = 0$, there is neither dilational nor radial displacement. The displacement at any point is directed tangentially at right angles to the radius drawn from the center of the sphere. The characteristic equation for the lth harmonic of the torsional oscillations is

$$(l-1)\Psi'_l(ka) + ka\Psi''_l(ka) = 0 \tag{5.6}$$

where a = radius of the sphere
ka = a quantity p/v_s or $(2\pi\omega)v_s$
$\Psi'_l(x)$ = a function expressible by

$$\Psi'_l(x) = \left(\frac{1}{x}\frac{d\Psi}{dx}\right)\left(\frac{\sin x}{x}\right)$$

or, in terms of a Bessel's function,

$$\Psi'_l(x) = (-1)^l \frac{\sqrt{2\pi}}{2} x^{-(l+1/2)} J_{(l+1/2)}(x) \tag{5.7}$$

Using Eq. (5.7), Eq. (5.6) becomes

$$(l-1)J_{(l+1/2)}(ka) - kaJ_{(l+3/2)}(ka) = 0 \tag{5.8}$$

If $l = 1$, we have rotary vibrations. In this case, Eq. (5.8) is given by

$$J_{5/2}(ka) = 0 \tag{5.9}$$

The lowest roots of this equation are

$$ka = 5.763, 9.093, 12.322\ldots$$

The actual modes corresponding to these solutions are given by $_2T_1$, $_3T_1$, $_4T_1$, and so on, as shown schematically in Fig. 5.2. The numerical

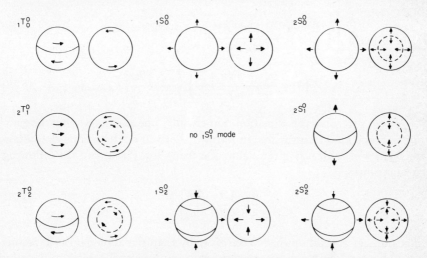

Fig. 5.2 Schematic displacements for T and S modes.

solutions for higher harmonic oscillations have been obtained by Sato and Usami.[5] The actual numbers are listed in Table 5.1.

TABLE 5.1 Nondimensional Frequency of Torsional Oscillation $_nT_l$ of Homogeneous Sphere

Values of l	$n=1$	$n=2$	$n=3$	$n=4$	$n=5$
1	......	5.7635	9.0950	12.3229	15.5146
2	2.5011	7.1360	10.5146	13.7717	16.9831
3	3.8647	8.4449	11.8817	15.1754	18.4121
4	5.0946	9.7126	13.2109	16.5445	19.8094
5	6.2658	10.9506	14.4108	17.8858	21.1806
6	7.4036	12.1664	15.7876	19.2042	22.5298
7	8.5199	13.3646	17.0453	20.5034	23.8601
8	9.6210	14.5484	18.2871	21.7861	
9	10.7109	15.7204	19.5152	23.0545	
10	11.7921	16.8821	20.7315	24.3104	

The spheroidal oscillation of a sphere involves both transverse and radial components of v_s and v_l. The displacement of a particle is given schematically in Fig. 5.2. By using a quantity $ha = p/v_l$, where v_l is the longitudinal sound wave velocity, the characteristic equation to be solved is in the form

$$\frac{2h}{k}\left[\frac{1}{ka} + \frac{(n-1)(n+2)}{(ka)^2}\frac{J_{(n-3/2)}ka}{J_{(n+3/2)}ka}\frac{n+1}{ka}\right]J_{(n+3/2)}(ha)$$

$$+ \left[-\frac{1}{2} + \frac{(n-1)(2n+1)}{(ka)^2} + \frac{1}{ka}\right]$$

$$\times \left(1 - \frac{2n(n-1)(n+2)}{(ka)^2}\right)\frac{J_{(n+3/2)}ka}{J_{(n+1/2)}ka}\right]J_{(n+1/2)}(ha) = 0 \quad (5.10)$$

In order to solve this equation numerically, one needs to know the ratio h/k. Since h/k is expressible in terms of Poisson's ratio σ, one can obtain the values of k for any given value of σ using Eq. (5.10) and the following relationship:

$$\frac{h}{k} = \left[\frac{1-2\sigma}{2(1-\sigma)}\right]^{1/2} \quad (5.11)$$

The numerical solutions have been given by Lamb[3] and by Sato and Usami[5] for a few values of σ. More extensive computations have been made by

Fraser and LeCraw,[1] who constructed a very useful figure showing the relationship between Poisson's ratio and the nondimensional frequencies of spheroidal oscillations. Their figure is reproduced here as Fig. 5.3.

The nondimensional frequencies of a few spheroidal oscillations are given in Table 5.2 as functions of Poisson's ratio.

In any event, the quantities π/ha and π/ka are the ratios of the period of oscillation to the time taken by dilatational or torsional waves to travel over a distance equal to the diameter of a sphere. Therefore, if one knows the resonant frequency f of a particular torsional mode, one can calculate the shear velocity v_s of the solid from the diameter d of a sphere by

$$v_s = \frac{\pi f d}{ka} \tag{5.12}$$

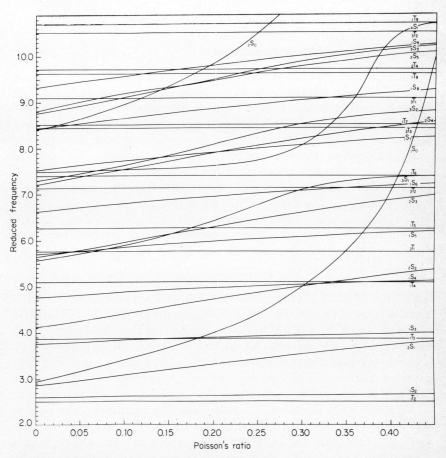

Fig. 5.3 Reduced frequency vs. Poisson's ratio. Graph used to simultaneously identify spherical resonance modes and determine Poisson's ratio.

TABLE 5.2 Nondimensional Frequency of Spheroidal Oscillation $_nS_l$ of Homogeneous Sphere

	Poisson's ratio				
Mode	0.10	0.17	0.25	0.33	0.40
$_1S_0$	3.3977	3.8071	4.4400	5.4322	7.0952
$_1S_2$	2.6152	2.6279	2.6399	2.6497	2.6569
$_1S_3$	3.8360	3.8771	3.9163	3.9489	3.9731
$_1S_4$	4.8747	4.9425	5.0093	5.0662	5.1090
$_1S_5$	5.8511	5.9416	6.0327	6.1118	6.1722
$_2S_1$	3.0737	3.2424	3.4245	3.5895	3.7168
$_2S_2$	4.4129	4.6263	4.8653	5.0878	5.2619
$_2S_3$	5.9597	6.1924	6.4544	6.6959	6.8811
$_3S_1$	5.8951	6.2383	6.7713	7.2306	7.3705

5.3 Specimens and Their Preparation

A suitable size for the measurement in question is a small sphere of about a few millimeters, but smaller sizes than this, say a 1-mm-diameter sphere, could be used, although it is not easy to fabricate or to handle experimentally, and spheres of only a few hundred microns in diameter have been measured successfully.

Several types of devices could be used to make spheres. A number of articles have appeared which describe methods of producing spherical specimens of small size. The apparatus falls into two categories: (1) contrarotating lapping pipes or turntables and (2) air-driven or motor-driven grinding cups. In both methods, sphericity is obtained only through random motion of the specimen. Unless a random grinding pattern is maintained, the finished piece will tend to be ellipsoidal, as noted by Durand.[6]

One of the pneumatic sphere grinders used in the authors' laboratory is shown in Fig. 5.4. The grinding cup, about 3 in in diameter, is a dense alumina ceramic crucible or a brass cup with a diamond lining. The latter can be used to make a sphere from very hard materials. A lid made of two lucite disks permits the entry of air and guides the air through six jets. The air is vented through the center of the disk and a tuned exit port provides control over the existing airflow. A pressure of 5 to 10 psi is sufficient to operate the apparatus. An asphericity of 0.2 percent may be achieved by this method. To decrease this asphericity, the contra-rotating lapping pipes used in lapidary work, or the turntables used for making precise balls for ballbearings, are recommended. With careful operation of these devices, an asphericity of less than 0.01 percent may be achieved.[7]

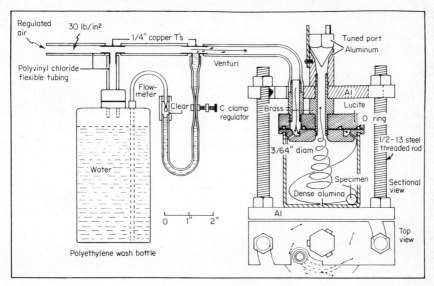

Fig. 5.4 Air-driven sphere grinder.

5.4 Equipment

A detailed description of the electronic arrangement is given by Fraser and LeCraw.[1] An electronic system for measuring the resonant frequencies of a sphere is shown in Fig. 5.5. The components are

1. Frequency counter—should be able to measure a frequency range 0 to 5 MHz with readout capacity of 1 cps on the 1-s interval. Hewlett-Packard Model 5244L or equivalent will serve.

2. Frequency synthesizer—should have continuous frequency control from about 100 kHz to at least 3 MHz. High-frequency stability and controllable output range to 2 V into 50-Ω resistive load are the basic requirements. A Rhode and Schwartz XUA synthesizer, General Radio coherent decade frequency synthesizer Type 1163A, or equivalent is preferred because of the ease of fine tuning to within a cycle.

3. Power amplifier—should have a maximum output of at least 2 W. A typical circuit for the amplifier is shown in Fig. 5.6.

4. Solid state switch—permits the transducer to be driven by the signal from the power amplifier and the return signal from the transducer to go to the receiving circuit when the one-transducer method is used. A single-pole, double-throw switch should have faster switching action and higher isolation than the conventional one. The model DS-16 Solid State Switch by Sanders Associates, or its equivalent, can be used without any modification.

134   Elastic Constants and Their Measurement

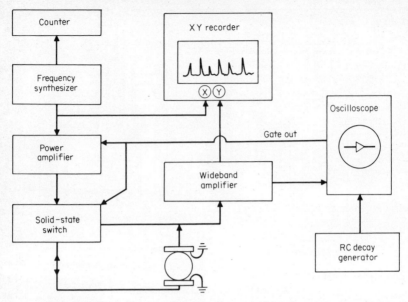

Fig. 5.5 Block diagram of electronic system for sphere resonance.

5. Amplifier—a wideband amplifier, such as the Hewlett-Packard Model 461A or Model 462A, or a tuned rf amplifier can be used.

6. Oscilloscope—should have a dual time base to provide the trigger signals for operation of the gated and switched components. The dual channel is required to display both the signal from the receiver and a variable RC decay curve for measuring Q of the specimen.

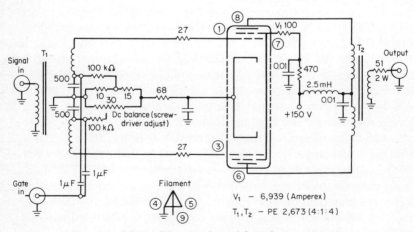

Fig. 5.6 Rf amplifier circuit used for sphere resonance.

7. *XY* recorder—serves to record an acoustic spectrum of the specimen by feeding the sweeping frequency into the x axis. This is also useful in determining Q from Δf (peak frequency width of a resonant vibration), but is not essential.

8. Transducers—most of the shear mode transducers whose resonant frequencies lie between 1 to 3 MHz, including PZT-5 and PZT-7, may be used. PZT-type transducers have been found to be adequate for measurements at temperatures lower than 100°C. For high temperatures, transducers having a high Curie-point temperature, such as of lithium niobate, have been successfully used.[8]

5.5 Method of Operation

The method of operation may be understood by referring to the diagrammatic sketch of Fig. 5.5.

The signal from the frequency source is fed into a power amplifier and is gated by using pulses from the oscilloscope. Each pulse coming out of the amplifier and entering the solid state switch contains a number of high-frequency waves generated by the synthesizer. The action of the switch is controlled by the signals from the oscilloscope. The pulse from the amplifier is fed to a shear mode transducer when the switch is ON; when the switch is OFF no signal is sent to the transducer. During this interval the transducer now acts as a receiver, detecting any signals caused by the free vibration of the sphere. These signals are amplified and displayed on the oscilloscope.

The detection of the resonant condition can be made by watching the pattern on the oscilloscope as the frequency is swept. When the frequency applied to the transducer is not equal to a free-oscillation frequency of the sphere, the sphere does not vibrate when the signal is cut off and no pattern appears during the gated OFF period (this is illustrated in Fig. 5.7). When the frequency of the transducer matches a resonant frequency of the sphere, the sphere continues to vibrate after the sending signal is cut off, so that a decay pattern appears on the oscilloscope screen. The logarithmic decay of this pattern contains information concerning the speed at which energy dies out and is related to the value of Q, so that the internal friction of a sphere can be calculated from this logarithmic decrement and the frequency of vibration. Among the numerous torsional modes possible, the $_2T_l$ modes are affected less by the "clamping" condition of the specimen than are other modes, possibly because the $_2T_l$ modes have nodal surfaces inside the sphere near its contact with the transducer, as shown in Fig. 5.2. For this reason, Fraser and LeCraw recommend the use of the $_2T_1$ mode for the determination of Q.

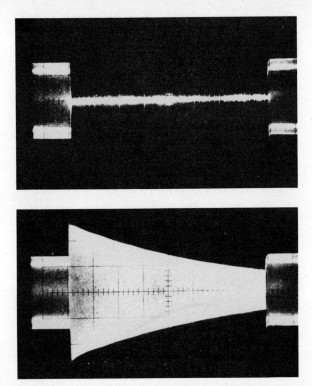

Fig. 5.7 Oscilloscope display showing nonresonant and resonant conditions. (*Upper*) Nonresonant condition: no pattern appears in the gated condition. (*Lower*) Resonant condition: a decay pattern appears.

5.6 Analysis of Data

The resonant-sphere technique has been applied to the determination of the elastic constants of several materials. One of the earliest applications was to tektites,[7] which are thought by some to be of extraterrestrial origin. Since they are usually so small in size, most other techniques could not be usefully applied to measure their elastic constants.

Figure 5.8 shows the acoustic spectrum obtained for several tektite samples. The peaks are distinct and sharp, and the identification of the proper mode is made easier if one bears the following points in mind. From the mathematical solution given in Sec. 5.2, the lowest possible mode is $_1T_2$, preceding a $_2S_1$. Although the resonant frequency of the $_2S_1$ mode depends on Poisson's ratio of the material being tested, its dependence is very small and the difference in frequency between $_1T_2$ and $_2S_1$ is about 5 percent. In general, $_1T_l$ modes show the lowest Q, S modes show medium-to-high Q, and $_2T_l$ modes, the highest Q. The first $_2T_l$

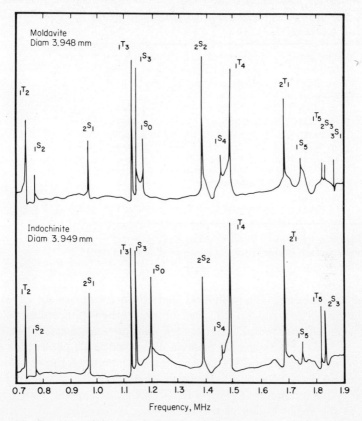

Fig. 5.8 Spectrum of the spherical modes of vibration of two tektite specimens.

mode is $_2T_1$, the nondimensional frequency of which is 5.7635, which is about 2.3 times as much as the $_1T_2$ mode.

From any $_nT_l$ mode, one can calculate the shear velocity by using Eq. (5.12). The velocity calculated from different modes should agree to within 0.2 percent for a normal specimen. An example is given for a tektite in Table 5.3, where the theoretical and experimental values for low modes of vibration are compared. Another example given is for polycrystalline spinel specimens, whose porosities range from 21 to 18 percent.[9] Although the acoustic quality of the specimen was poor due to the high porosity, the agreement of the values of v_s calculated from the various modes is very good, as shown in Table 5.4.

When the specimen is not spherical but ellipsoidal, resonant frequencies split into several modes, depending upon the types of oscillation. This problem was solved mathematically by Usami and Sato[10] for torsional oscillation of a homogeneous elastic spheroid with regard to problems

TABLE 5.3 Resonant Frequencies of an Indochinite Sphere (Diameter, 0.6081 cm)

Mode	Observed frequency f_i, MHz	Observed nondimensional frequency*	Theoretical nondimensional frequency†
$_2S_1$	0.63608	3.3329	3.3307
$_1T_3$	0.73725	3.8635	3.8647
$_1S_3$	0.74380	3.8978	3.8967
$_1S_0$	0.78186	4.0972	4.0788
$_2S_2$	0.90512	4.7432	4.7412
$_1S_4$	0.95045	4.9807	4.9756
$_1T_4$	0.97215	5.0944	5.0946
$_2T_1$	1.09983	5.7635	5.7635

* $5.7635 \times f_i/f_{2T_1}$.
† Solution of the characteristic equations for Poisson's ratio of 0.2080.

related to the earth. They showed that ellipticity causes the $_nT_l$ modes to split into $n+1$ modes, all close to each other. According to their calculation, ellipticity of 0.3 percent (or the difference between the major and minor axis of 0.3 percent) causes a difference of 0.7 percent for $l=2$, 0.5 percent for $l=3$, and 0.4 percent for $l=4$ between the extreme frequencies for each value of n. However, they noted that a spheroid has a frequency for the mode $_1T_l$ nearly equal to that of a sphere with a radius equal to the mean radius of the spheroid. It has been found through several experiments that the variation of the frequency is less than one-quarter of the variation of the diameter for the $_2T_1$ mode of vibration. Thus, the accuracy of the sound velocity data depends mainly on how accurately the specimen is fabricated.

Poisson's ratio σ is found through awareness that the $_nT_l$ modes are not dependent on that ratio, whereas the $_nS_l$ modes are dependent on it as well

TABLE 5.4 Results for Spinel Specimens

Specimen	Shear velocity, km/s, spheroidal mode used								Poisson's ratio
	$_1T_2$	$_1T_3$	$_1T_4$	$_2T_1$	$_1T_5$	$_2T_2$	$_1T_6$	Average	
1	4.737	4.742	4.731	4.757	4.742	4.746	4.744	4.742	0.244
2	4.671	4.686	4.679	4.686	4.694	4.689	4.682	4.690	0.247
3	4.879	4.891	....	4.888	4.891	4.884	4.890	4.890	0.245
4	4.923	4.927	4.929	4.929	4.925	4.919	5.926	4.926	0.245
5	4.928	4.923	4.920	4.926	4.917	4.921	4.916	4.921	0.249

as on the shear sound velocity. Therefore, when the frequency ratio of, say, the $_2T_1$ mode to the $_1S_0$ mode is $5.7635/4.4400 = 1.2980$, Poisson's ratio of the materials being determined is 0.250. By using Fig. 5.3, one may estimate Poisson's ratio within two to three significant figures. In using Fig. 5.3, one first compiles a list of the observed resonant frequencies, in either ascending or descending order. These are then converted to reduced frequencies. This is accomplished in the following fashion. Suppose one assumes that the lowest frequency observed is due to the $_1T_2$ mode, and that this assumption is supported by the fact that it was also a low-Q mode. (If the lowest frequency were a high-Q mode, one would guess it might be due to the $_2S_1$ mode; the highest Q modes are generally $_2T_1, {_2S_1}, {_2S_0}$.) By assigning this lowest frequency as the $_1T_2$ mode we have assumed that the nondimensional (reduced) frequency is 2.5011 (see Table 5.1). We now construct the corresponding list of reduced frequencies by multiplying each frequency by the ratio 2.5011 to the frequency of the $_1T_2$ mode, that is, $2.5011/f_{1T_2}$. This computation yields a list of the nondimensional (reduced) frequencies appropriate to each mode. For the T modes, these would be values corresponding to those in Table 5.1, while for the S modes, Table 5.2 should be used if the specimen has a value of Poisson's ratio listed there. If the specimen were to have a Poisson's ratio of say 0.17, then we would have a reduced frequency table of values such as the following.

Measured frequency spectrum, kHz	Reduced frequency, calculated	Reduced frequency theoretical (Table 5.1 or 5.2)	Assigned mode
0.3515	2.5011	2.5011	$_1T_2$
0.3693	2.628	2.6279	$_1S_2$
0.4559	3.244	3.2424	$_2S_1$
0.5354	3.810	3.8071	$_1S_0$
0.5429	3.863	3.8647	$_1T_3$
0.5449	3.877	3.8771	$_1S_3$
0.6499	4.625	4.6263	$_2S_2$
0.6948	4.944	4.9425	$_1S_4$
0.7160	5.095	5.0946	$_1T_4$

The values of the $_nT_l$ modes are independent of Poisson's ratio, but the values of the $_nS_l$ modes are not. As a consequence, the $_nS_l$ mode assigned to a reduced frequency will be different for different values of Poisson's ratio, and this may result in a change in the order of assigning modes to the reduced frequencies. For example, if Poisson's ratio were 0.20 in the above example, we would have an interchange in the order of modes from ... $_2S_1$, $_1S_0$, $_1T_3$, ... to ... $_2S_1$, $_1T_3$, $_1S_3$, $_1S_0$ ..., whereas if Poisson's ratio were 0.1, then the order of $_1T_3$, $_1S_3$ would interchange. To interpret the spectrum for a specimen of unknown Poisson's ratio, a strip of paper is placed against the vertical scale of Fig. 5.3, and the calculated reduced frequencies are marked off to scale. The strip of paper is then translated *horizontally* until the reduced frequencies marked off intersect the modes plotted, to yield the best fit. Poisson's ratio is read directly from the figure, and the modes may then be assigned, the appropriate mode corresponding to the intersection with the marked-off reduced frequency. If a fit cannot be attained, then the assumption that the lowest mode was the $_1T_2$ was wrong; a new mode is assigned and the process repeated. Because the T modes are independent of Poisson's ratio, it is preferable to attempt to identify one of the tabulated frequencies as corresponding to a T mode, and using it to assign the reduced frequency values.

Once the modes have been properly identified and assigned, Poisson's ratio is immediately determined to three significant figures, and the shear velocity can be calculated by using Eq. (5.12). Poisson's ratio depends very sensitively upon the assignment of the torsional and spheroidal modes—and so this property may be precisely determined from the data. For example, in Fig. 5.8 we see that the difference in frequency between the $_1T_3$ and $_1S_0$ modes for indochinite is larger than that for moldavite. This indicates that Poisson's ratio for indochinite is higher than that for moldavite. A precise value can be determined by finding the value of h/k which satisfies the experimentally determined frequency ratio using Eq. (5.10) and a computer.

The calculation of Poisson's ratio from the $_2T_1$ and $_1S_0$ modes, using the characteristic equations, requires only the frequency ratio of these two modes. Since error in the frequency ratio is in the order of 0.01 percent, the error in Poisson's ratio is also in the order of 0.01 percent. For this reason, the resonant-sphere technique is most suitable for determining Poisson's ratio, and one may expect to detect a variation in that ratio as small as 0.0001. An example is given in Fig. 5.9; it shows the porosity dependence of Poisson's ratio for polycrystalline MgO.[11] It is clear that Poisson's ratio increases with increasing porosity for MgO.

The development of high-temperature transducer materials, such as lithium niobate, permits application of the technique in determining the sound velocity at high temperatures—an example is the measurements on alumina to temperatures up to 1000°C.[8] The high resolution of the data obviates the need for a wide temperature range in order to determine the temperature derivatives of sound velocities, if one is only interested in the value near room temperature.

The problems related to the anisotropic nature of the specimen have not

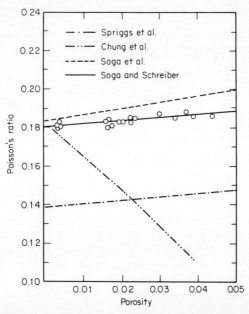

Fig. 5.9 Poisson's ratio of polycrystalline MgO as a function of porosity. [*Data from Spriggs et al., J. Am. Ceram. Soc.*, **45**: 400 (1962); *Chung et al., ibid.*, **46**: 452 (1963); *Soga et al., ibid.*, **49**: 318 (1966); *Soga and Schreiber, ibid.*, **51**: 465 (1968).]

yet been solved, although it has been noted that a resonance peak splits into several peaks depending on the mode and the degree of anisotropy of the specimen.

REFERENCES

1. Fraser, D. B., and R. C. Lecraw: Novel Method of Measuring Elastic and Anelastic Properties of Solids, *Rev. Sci. Instr.*, **35**:113 (1964).
2. Kelvin, Lord: On the Rigidity of the Earth, *Philos. Trans. Roy. Soc.*, **153**:573 (1863).
3. Lamb, H.: On the Vibrations of an Elastic Sphere, *Proc. Math. Soc.*, **13**:189 (1882).
4. Love, A. E. H.: "A Treatise on the Mathematical Theory of Elasticity," Dover, New York, 1944.
5. Sato, Y., and T. Usami: Basic Study on the Oscillation of Homogeneous Elastic Sphere, *Geophys. Mag.* (Japan Meteorological Agency, Tokyo), **31**:15 (1962).
6. Durand, J.: Making Spheres of Crystals with Anisotropy of Hardness, *Rev. Sci. Instr.*, **30**:840 (1959).
7. Soga, N., and O. L. Anderson: Elastic Properties of Tektites Measured by Resonant Sphere Technique, *J. Geophys. Res.*, **72**:1733 (1967).
8. Fraser, D. B., and A. Q. Warner: Lithium Niobate: A High-Temperature Piezoelectric Transducer Material, *J. Appl. Phys.*, **37**:3853 (1966).
9. Schreiber, E.: Comment upon the Elastic Modulus Porosity Relationship. *J. Am. Ceram. Soc.*, **51**:541 (1968).

10. Usami, T., and Y. Sato: Torsional Oscillation of a Homogeneous Elastic Spheroid, *Bull. Seismol. Soc. Am.*, **52**:469 (1962).
11. Soga, N., and E. Schreiber: Porosity Dependence of Sound Velocity and Poisson's Ratio for Polycrystalline MgO Determined by Resonant Sphere Method, *J. Am. Ceram. Soc.*, **51**:465 (1968).

CHAPTER SIX

Indirect Methods of Estimating Elastic Constants

6.1 Introduction

The preceding chapters have been concerned with the precise experimental techniques used to determine elastic constants, or sound velocities, on small specimens. The experimentalist who is designing apparatus will find this treatment of interest. In some circumstances, however, it is useful to be able to estimate the magnitude of elastic moduli without the necessity of measurement. In this chapter, several methods of estimating the magnitude of such moduli are presented.

The methods described all rest upon the principle that elastic properties are a measure of the forces between atoms. There are other physical properties which also depend upon such forces, and, in principle, it should be possible to relate the elastic properties to the other physical properties. Examples of the latter include specific heat, infrared optical spectra, and compressibility. These are properties of solids which are of great interest, and data on one or more of these properties are usually available. For this reason, it is desirable to explore the relationships from which the elastic moduli may be estimated. The equations presented in this chapter involve approximations in theory which are often crude, and they should not be used when direct experimental data concerning elastic constants are available.

6.2 Estimating Elastic Constants

6.2.1 Estimating the Shear Velocity from the Specific Heat.

The isotropic shear velocity v_s of an inorganic material such as a mineral or rock can be estimated from its low-temperature specific heat.[1] This approach promises to be very useful, because one can determine the velocity with which shear waves propagate through a material independently of the state of aggregation of the samples. This correspondence between acoustics and calorimetry is based upon a principle of lattice dynamics which states that at sufficiently low temperatures the optical vibrations of a solid are quiescent and the vibrational energy arises solely from acoustic vibrations. The correspondence is conveniently stated in terms of the Debye temperatures.

Low-temperature specific heat is represented by a scalar parameter called the thermal Debye temperature θ_t, and the acoustic specific heat is represented by the acoustic Debye temperature θ_a. Thus, at temperatures near absolute zero,

$$\theta_t = \theta_a \tag{6.1}$$

The expression for θ_a in terms of the sound velocities for an isotropic body is given by

$$\theta_a = \frac{h}{k}\left[\frac{9\rho N}{4\pi(M/p)}\right]^{1/3}\left(\frac{2}{v_s^3}+\frac{1}{v_l^3}\right)^{-1/3} \tag{6.2}$$

where h, k, and N are Plank's constant, the Boltzman constant, and Avogadro's number, respectively, M/p is the mean atomic weight (the molecular weight divided by the number of atoms p involved in the molecular formula), ρ is the density, and v_s and v_l are the shear and compressional velocities. It is more convenient to write this expression in terms of the mean sound velocity v_m:

$$\frac{3}{v_m^3}=\frac{2}{v_s^3}+\frac{1}{v_l^3} \tag{6.3}$$

Combining the preceding equations results in

$$\theta_t = 251.4\left(\frac{p\rho}{M}\right)^{1/3} v_m \tag{6.4}$$

Here the units of v_m are kilometers per second, and the numerical factor arises from the physical and numerical constants in the preceding equations.

Indirect Methods of Estimating Elastic Constants

It may be observed that both v_l and v_s are needed for defining θ_t. A closer examination, however, reveals that as a good approximation v_s alone defines θ_t.

We first solve Eq. (6.3) for the ratio of v_s/v_m.

$$\frac{v_s}{v_m} = \left[\frac{2}{3} + \frac{1}{3}\left(\frac{v_s}{v_l}\right)^3\right]^{1/3} \quad (6.5)$$

From the relation between Poisson's ratio σ and v_s and v_l (Table 1.1),

$$\sigma = \frac{1}{2}\left\{1 - \left[\left(\frac{v_l}{v_s}\right)^2 - 1\right]^{-1}\right\} \quad (6.6)$$

and solving for the ratio of v_s/v_l, we obtain

$$\frac{v_s}{v_l} = \left[\frac{1-2\sigma}{2(1-\sigma)}\right]^{1/2}$$

which is substituted into Eq. (6.5) and solved, yielding

$$\frac{v_s}{v_m} = \left\{\frac{2}{3} + \frac{1}{3}\left[\frac{1-2\sigma}{2(1-\sigma)}\right]^{3/2}\right\}^{1/3} \quad (6.7)$$

The ratio v_s/v_m is a slowly varying function of Poisson's ratio, as shown in Fig. 6.1. This ratio has the value 0.900 at $\sigma = 0.25$, and lies within

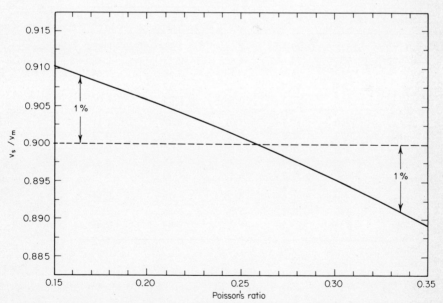

Fig. 6.1 Range of Poisson's ratio, over which deviation of values are less than 1%. From assumption in Eq. (6.8).

1 percent of that value over the interval $0.15 < \sigma < 0.35$. Thus, within the limits of a 1 percent error, we can write for Eq. (6.7)

$$v_s = 0.9\, v_m \tag{6.8}$$

It is this lack of sensitivity of the value of the ratio v_s/v_m to a wide range of values of Poisson's ratio that makes this approximation for estimating the shear velocity a useful one. If we now substitute from Eq. (6.8) into Eq. (6.4) and solve for v_s, we arrive at the final form of the equation for estimating v_s from a knowledge of the Debye temperature. This is

$$v_s = \frac{\theta_t}{279.4}\left(\frac{M}{p\rho}\right)^{1/3} \tag{6.9}$$

This equation has many practical uses. A knowledge of the mean atomic weight, density, and low-temperature specific heat (from which θ_t can be obtained) is sufficient for computing v_s. Some examples of the use of this equation are shown in Table (6.1). For these cases, the agreement is generally 1 or 2 percent.

TABLE 6.1 Shear Velocity Calculated from Eq. (6.9)

Material	θ_t, °K	M/p	ρ, g/cm³	v_s, km/s Calculated	v_s, km/s Measured
NaCl	321	29.2	2.16	2.73	2.62
KCl	235	37.3	1.99	2.23	2.19
CaF$_2$	510	26.0	3.18	3.68	3.64
Zn	315	48.7	4.08	2.57	2.83
MgO	946	20.3	3.58	6.03	6.04
BeO	1200	12.5	3.01	6.87	7.11
α − SiO$_2$	572	20.0	2.65	4.02	4.13
TiO$_2$	760	26.6	4.26	5.00	5.15
α − Al$_2$O$_3$	1045	20.4	3.99	6.45	6.43
Fe$_2$O$_3$	660	31.9	5.27	4.30	4.16
Al	428	26.9	2.70	3.29	3.13
Fe	445	55.8	7.87	3.06	3.20
W	184	183.9	19.2	2.89	2.86

Discrepancies between the measured value of v_s and the value calculated from Eq. (6.9) can arise in three ways. First, Eq. (6.1) may not be completely valid. This question has been discussed by Alers,[2] who found that it holds for crystalline nonmetals; but may not do so for metals, because of the electronic contribution to the specific heat which does not contribute to the lattice modes governing sound propagation. Second, the approximation $v_s = 0.9 v_m$ will not hold for materials with a very low Poisson's ratio. The value of σ for BeO is 0.11, and there is a corresponding lack of agreement for v_s. Third, the value of θ_t derived from specific-heat data may be incorrectly determined, and the method of finding θ_t from the specific heat recommended by Barron et al.[3] is preferred. Specific-heat measurements at liquid nitrogen temperature or higher temperatures are often reported. Values of C_p in this temperature range will probably lead to poor estimates of v_s made with Eq. (6.9), since C_p may then be influenced by optical vibrations in addition to the acoustic vibrations.

From the results of Alers[2] it appears that Eq. (6.5) is valid for all crystalline materials of interest in ceramics and petrology. It may, however, not apply to inorganic glasses. It has been shown not to hold for vitreous silica[4] though it does hold for quartz.

It should be possible to calculate v_s from suitable low-temperature, specific-heat measurements with all manner of specimens: rocks, crystals, sediments, dust, pumice, and aggregates. The resulting value will be independent of microstructure effects arising from grain boundaries and pores and will correspond to the dense isotropic solid at zero porosity.

Equation (6.9) is as accurate as many standard techniques for measuring the v_s of rocks and coarse aggregates, where the determination of the v_s of aggregates by resonance or acoustic techniques may yield values that are uncertain because of the presence of pores, internal strains, or anisotropy.

6.2.2 Using the Debye Temperature and Bulk Modulus to Find the Isotropic Elastic Moduli. The method described in Sec. 6.2.1 above, is helpful when the Debye temperature is the only available datum. Very often, when sound velocities are not available, the literature contains either the bulk modulus or Young's modulus in addition to the specific-heat data. The prevalence of the latter data results from the fact that the bulk modulus, which is also a scalar quantity, can be determined for materials that are in granular or crystalline form. Young's modulus, which is important as a criterion in the design of stressed components, is also often available by itself for many materials. When both the Debye temperature and either the bulk or Young's modulus are available, it is possible to circumvent the approximation used above ($v_s = 0.9 v_m$), and a method is available to permit the calculation of both the shear and longitudinal velocities. This is accomplished in the following manner.[5]

The starting points are Eqs. (6.3) and (6.4), which define the mean sound velocity and the Debye temperature. We can also express the mean sound velocity in terms of any pair of elastic moduli according to the equations given in Table (1.1). In particular it is convenient to use either the bulk modulus B or Young's modulus E together with Poisson's ratio. The pertinent equations are

$$\rho v_l^2 = 3B \frac{1-\sigma}{1+\sigma} = \frac{E(1-\sigma)}{(1+\sigma)(1-2\sigma)}$$

$$\rho v_s^2 = 3B \frac{1-2\sigma}{2+2\sigma} = \frac{E}{2+2\sigma}$$

(6.10)

and substituting these into Eq. (6.3) and solving for v_m, yields the corresponding pair of equations,

$$v_m = \left\{ \frac{3}{2[(2+2\sigma)/3(1-2\sigma)]^{3/2} + [(1+\sigma)/3(1-\sigma)]^{3/2}} \right\}^{1/3} \left(\frac{B}{\rho}\right)^{1/2}$$

(6.11)

and $$v_m = \left\{ \frac{3}{[2(2+2\sigma)]^{3/2} + [(1+\sigma)(1-2\sigma)/(1-\sigma)]^{3/2}} \right\}^{1/3} \left(\frac{E}{\rho}\right)^{1/2}$$

We can now obtain a useful form of Eq. (6.4) by combining either of Eqs. (6.11) with it. Taking the one containing the bulk modulus and Poisson's ratio, and rearranging the terms, we obtain an equation in dimensionless form; thus:

$$Z_B = \frac{k}{h} \left(\frac{3p\rho}{M}\right)^{-1/3} \left(\frac{B}{\rho}\right)^{-1/2} \theta_t$$

$$= \left\{ \frac{3}{[2(2+2\sigma)/3(1-2\sigma)]^{3/2} + [(1+\sigma)/3(1-\sigma)]^{3/2}} \right\}^{1/3}$$

and (6.12)

$$Z_E = \frac{k}{h} \left(\frac{3p\rho}{M}\right)^{-1/3} \left(\frac{E}{\rho}\right)^{-1/2} \theta_t = \left\{ \frac{3}{2(2+2\sigma)^{3/2} + [(1+\sigma)(1-2\sigma)/(1-\sigma)]^{3/2}} \right\}^{1/3}$$

where Z_B and Z_E are appropriate dimensionless parameters.

Suppose θ_t and B are known for a particular material, then Z_B may be calculated and σ determined from the right-hand term. Now knowing both B and σ, v_s and v_l may be calculated from Eq. (6.10), and a similar procedure is followed if θ_t and E are known. For convenience in converting Z_B or Z_E to values of σ, use may be made of Tables 6.2 and 6.3, from which Poisson's ratio may be read for different values of Z.

As an example, the compressibility as measured on a powdered sample of MgO is reported to be 0.60×10^{-6} atm^{-1}, which corresponds to a bulk modulus of 1,680 kbar.[6] The density of MgO is 3.583 and the Debye temperature is 946°K (see Table 6.1). From Eqs. (6.12), Z is found to be 0.9829, and this yields a Poisson's ratio of 0.186 (Table 6.2). The shear and longitudinal velocities calculated from these values of B and σ are 9.825 and 6.102 km/s, respectively, which compare rather well with the measured values. It is worth pointing out that this method yields a knowledge of two elastic moduli, either B and σ or E and σ. It is therefore possible to calculate any of the isotropic elastic moduli using the relations given in Table 1.1, so this approach is not limited to the estimation of velocity only, as is the method described in Sec. 6.2.1 above.

The method of computing elastic moduli from the Debye temperature and bulk modulus was apparently first indicated by Potter.[7] It is a reversal of the well-known procedure of making the Debye temperature an explicit function of elastic constants, which is a standard approach in the elementary treatment of lattice dynamics.

6.3 Estimation of the Bulk Modulus from the Volume

6.3.1 Pure Compounds. The bulk modulus at constant temperature is a function of volume only. The volume of a solid will depend upon the attractive and repulsive forces holding the solid together, so that at the temperature of absolute zero where there are no thermal contributions to the volume, the volume of the solid will be determined entirely by the nature of the interionic forces. In this section, we will develop equations relating the bulk modulus to the volume following this approach. The development employed will be based upon the Born model of the potential for an ionic solid. In this model, the attractive forces are considered to arise from the electrostatic charges on the ions themselves, and the repulsive potential is assumed to follow a power law. The potential takes the form

$$U = -\frac{A Z_c Z_a e^2}{r} + \frac{b}{r^n} \qquad (6.13)$$

where Z_c and Z_a = charges of cation and anion, respectively
e = charge of the electron
r = ion separation
n = a constant
A = Madelung constant
b = repulsive parameter which remains to be determined

The Madelung constant arises from summing the attractive potential over distant neighbors of the ions composing the crystal, and therefore depends upon the spatial distribution of the ions. For this reason the constant is a structure-sensitive property. The Madelung constants for several structure types are listed in Table 6.4.

150 Elastic Constants and Their Measurement

TABLE 6.2 Conversion Table for Z_B and Poisson's Ratio

Poisson's ratio	0.000	0.001	0.002	0.003	0.004
0.000	1.32793	1.32604	1.32415	1.32226	1.32037
0.010	1.30905	1.30717	1.30529	1.30340	1.30152
0.020	1.29025	1.28838	1.28650	1.28463	1.28275
0.030	1.27153	1.26966	1.26779	1.26592	1.26405
0.040	1.25286	1.25100	1.24914	1.24728	1.24542
0.050	1.23426	1.23240	1.23054	1.22869	1.22683
0.060	1.21570	1.21385	1.21200	1.21014	1.20829
0.070	1.19719	1.19534	1.19349	1.19164	1.18979
0.080	1.17870	1.17686	1.17501	1.17316	1.17132
0.090	1.16024	1.15840	1.15655	1.15471	1.15286
0.100	1.14180	1.13996	1.13811	1.13627	1.13443
0.110	1.12336	1.12152	1.11968	1.11783	1.11599
0.120	1.10492	1.10308	1.10123	1.09939	1.09754
0.130	1.08647	1.08462	1.08278	1.08093	1.07908
0.140	1.06800	1.06615	1.06430	1.06245	1.06060
0.150	1.04949	1.04764	1.04578	1.04393	1.04208
0.160	1.03094	1.02908	1.02722	1.02537	1.02351
0.170	1.01234	1.01047	1.00861	1.00674	1.00488
0.180	0.99367	0.99180	0.98992	0.98805	0.98618
0.190	0.97492	0.97304	0.97116	0.96928	0.96739
0.200	0.95608	0.95419	0.95230	0.95040	0.94851
0.210	0.93713	0.93523	0.93332	0.93142	0.92952
0.220	0.91806	0.91614	0.91423	0.91231	0.91039
0.230	0.89885	0.89692	0.89499	0.89306	0.89113
0.240	0.87949	0.87754	0.87560	0.87365	0.87170
0.250	0.85995	0.85799	0.85602	0.85405	0.85208
0.260	0.84022	0.83823	0.83625	0.83426	0.83226
0.270	0.82027	0.81826	0.81625	0.81423	0.81222
0.280	0.80007	0.79804	0.79600	0.79396	0.79192
0.290	0.77961	0.77754	0.77548	0.77341	0.77134
0.300	0.75884	0.75675	0.75465	0.75255	0.75044
0.310	0.73774	0.73561	0.73348	0.73134	0.72920
0.320	0.71626	0.71410	0.71192	0.70974	0.70756
0.330	0.69437	0.69216	0.68994	0.68772	0.68549
0.340	0.67201	0.66975	0.66748	0.66521	0.66293
0.350	0.64913	0.64681	0.64448	0.64215	0.63982
0.360	0.62566	0.62327	0.62088	0.61849	0.61608
0.370	0.60151	0.59906	0.59660	0.59413	0.59165
0.380	0.57661	0.57407	0.57153	0.56897	0.56641
0.390	0.55083	0.54820	0.54555	0.54290	0.54024
0.400	0.52403	0.52129	0.51854	0.51577	0.51300
0.410	0.49606	0.49318	0.49030	0.48740	0.48448
0.420	0.46667	0.46364	0.46060	0.45754	0.45446
0.430	0.43558	0.43237	0.42913	0.42587	0.42259
0.440	0.40240	0.39894	0.39546	0.39195	0.38841
0.450	0.36653	0.36277	0.35897	0.35513	0.35125
0.460	0.32711	0.32292	0.31868	0.31439	0.31004
0.470	0.28264	0.27782	0.27292	0.26794	0.26287
0.480	0.23023	0.22434	0.21831	0.21210	0.20572
0.490	0.16240	0.15403	0.14519	0.13578	1.12568

0.005	0.006	0.007	0.008	0.009
1.31848	1.31659	1.31471	1.31282	1.31094
1.29964	1.29776	1.29588	1.29401	1.29213
1.28088	1.27901	1.27714	1.27527	1.27340
1.26219	1.26032	1.25846	1.25659	1.25473
1.24355	1.24169	1.23983	1.23798	1.23612
1.22497	1.22312	1.22126	1.21941	1.21756
1.20644	1.20459	1.20274	1.20089	1.19904
1.18794	1.18609	1.18425	1.18240	1.18055
1.16947	1.16763	1.61578	1.16393	1.16209
1.15102	1.14918	1.14733	1.14549	1.14364
1.13258	1.13074	1.12889	1.12705	1.12521
1.11414	1.11230	1.11046	1.10861	1.10677
1.09570	1.09385	1.09201	1.09016	1.08832
1.07724	1.07539	1.07354	1.07169	1.06985
1.05875	1.05690	1.05505	1.05319	1.05134
1.04022	1.03837	1.03651	1.03465	1.03280
1.02165	1.01979	1.01792	1.01606	1.01420
1.00301	1.00114	0.99928	0.99741	0.99554
0.98430	0.98243	0.98055	0.97867	0.97680
0.96551	0.96362	0.96174	0.95985	0.95796
0.94662	0.94472	0.94282	0.94093	0.93903
0.92761	0.92570	0.92379	0.92188	0.91997
0.90847	0.90655	0.90463	0.90270	0.90078
0.88919	0.88725	0.88531	0.88337	0.88143
0.86974	0.86779	0.86583	0.86387	0.86191
0.85011	0.84814	0.84616	0.84418	0.84220
0.83027	0.82827	0.82628	0.82428	0.82227
0.81020	0.80818	0.80616	0.80413	0.80210
0.78988	0.78783	0.78578	0.78372	0.78167
0.76926	0.76719	0.76510	0.76302	0.76093
0.74833	0.74622	0.74411	0.74199	0.73987
0.72705	0.72490	0.72275	0.72059	0.71843
0.70537	0.70318	0.70099	0.69879	0.69658
0.68325	0.68102	0.67877	0.67653	0.67427
0.66064	0.65835	0.65605	0.65375	0.65144
0.63747	0.63512	0.63276	0.63040	0.62803
0.61367	0.61126	0.60883	0.60640	0.60396
0.58916	0.58667	0.58417	0.58165	0.57913
0.56383	0.56125	0.55866	0.55606	0.55345
0.53757	0.53488	0.53219	0.52948	0.52676
0.51021	0.50740	0.50459	0.50176	0.49891
0.48156	0.47861	0.47565	0.47267	0.46968
0.45136	0.44824	0.44511	0.44195	0.43878
0.41928	0.41595	0.41260	0.40923	0.40582
0.38484	0.38125	0.37762	0.37395	0.37026
0.34734	0.34338	0.33938	0.33533	0.33124
0.30563	0.30117	0.29664	0.29204	0.28737
0.25771	0.25244	0.24707	0.24158	0.23597
0.19914	0.19234	0.18530	0.17799	0.17037
0.11471	0.10258	0.08883	0.07255	0.05136

TABLE 6.3 Conversion Table for Z_E and Poisson's Ratio

Poisson's ratio	0.000	0.001	0.002	0.003	0.004
0.000	0.76668	0.76636	0.76603	0.76571	0.76538
0.010	0.76346	0.76314	0.76282	0.76250	0.76218
0.020	0.76029	0.75998	0.75966	0.75935	0.75904
0.030	0.75718	0.75688	0.75657	0.75626	0.75596
0.040	0.75414	0.75384	0.75353	0.75323	0.75293
0.050	0.75115	0.75085	0.75056	0.75026	0.74997
0.060	0.74821	0.74792	0.74763	0.74735	0.74706
0.070	0.74534	0.74505	0.74477	0.74448	0.74420
0.080	0.74251	0.74224	0.74196	0.74168	0.74140
0.090	0.73975	0.73947	0.73920	0.73893	0.73865
0.100	0.73703	0.73676	0.73649	0.73623	0.73596
0.110	0.73437	0.73410	0.73384	0.73358	0.73331
0.120	0.73175	0.73150	0.73124	0.73098	0.73072
0.130	0.72919	0.72894	0.72869	0.72843	0.72818
0.140	0.72668	0.72643	0.72618	0.72594	0.72569
0.150	0.72422	0.72397	0.72373	0.72349	0.72325
0.160	0.72180	0.72156	0.72133	0.72109	0.72085
0.170	0.71944	0.71920	0.71897	0.71873	0.71850
0.180	0.71712	0.71689	0.71666	0.71643	0.71620
0.190	0.71484	0.71462	0.71439	0.71417	0.71395
0.200	0.71262	0.71240	0.71218	0.71196	0.71174
0.210	0.71043	0.71022	0.71000	0.70979	0.70957
0.220	0.70830	0.70808	0.70787	0.70766	0.70745
0.230	0.70620	0.70600	0.70579	0.70558	0.70538
0.240	0.70415	0.70395	0.70375	0.70354	0.70334
0.250	0.70214	0.70195	0.70175	0.70155	0.70135
0.260	0.70018	0.69998	0.69979	0.69960	0.69940
0.270	0.69825	0.69806	0.69787	0.69768	0.69749
0.280	0.69637	0.69618	0.69600	0.69581	0.69563
0.290	0.69452	0.69434	0.69416	0.69398	0.69380
0.300	0.69272	0.69254	0.69236	0.69218	0.69201
0.310	0.69095	0.69078	0.69060	0.69043	0.69025
0.320	0.68922	0.68905	0.68888	0.68871	0.68854
0.330	0.68752	0.68736	0.68719	0.68702	0.68686
0.340	0.68586	0.68570	0.68553	0.68537	0.68521
0.350	0.68423	0.68407	0.68391	0.68375	0.68359
0.360	0.68264	0.68248	0.68232	0.68216	0.68201
0.370	0.68107	0.68091	0.68076	0.68060	0.68045
0.380	0.67953	0.67937	0.67922	0.67907	0.67892
0.390	0.67801	0.67786	0.67771	0.67756	0.67741
0.400	0.67651	0.67636	0.67621	0.67606	0.67592
0.410	0.67503	0.67488	0.67473	0.67459	0.67444
0.420	0.67356	0.67341	0.67327	0.67312	0.67297
0.430	0.67210	0.67195	0.67180	0.67166	0.67151
0.440	0.67063	0.67048	0.67034	0.67019	0.67004
0.450	0.66915	0.66901	0.66886	0.66871	0.66856
0.460	0.66766	0.66750	0.66735	0.66720	0.66705
0.470	0.66612	0.66596	0.66580	0.66565	0.66549
0.480	0.66452	0.66435	0.66419	0.66402	0.66385
0.490	0.66281	0.66263	0.66245	0.66227	0.66208

0.005	0.006	0.007	0.008	0.009
0.76506	0.76474	0.76442	0.76410	0.76378
0.76187	0.76155	0.76123	0.76092	0.76060
0.75873	0.75842	0.75811	0.75780	0.75749
0.75565	0.75535	0.75505	0.75474	0.75444
0.75264	0.75234	0.75204	0.75174	0.75144
0.74967	0.74938	0.74909	0.74880	0.74851
0.74677	0.74648	0.74619	0.74591	0.74562
0.74392	0.74364	0.74336	0.74307	0.74279
0.74112	0.74085	0.74057	0.74030	0.74002
0.73838	0.73811	0.73784	0.73757	0.73730
0.73569	0.73543	0.73516	0.73489	0.73463
0.73305	0.73279	0.73253	0.73227	0.73201
0.73047	0.73021	0.72996	0.72970	0.72945
0.72793	0.72768	0.72743	0.72718	0.72693
0.72544	0.72520	0.72495	0.72471	0.72446
0.72300	0.72276	0.72252	0.72228	0.72204
0.72061	0.72038	0.72014	0.71991	0.71967
0.71827	0.71804	0.71781	0.71758	0.71735
0.71597	0.71575	0.71552	0.71529	0.71507
0.71372	0.71350	0.71328	0.71306	0.71284
0.71152	0.71130	0.71108	0.71087	0.71065
0.70936	0.70915	0.70893	0.70872	0.70851
0.70724	0.70703	0.70683	0.70662	0.70641
0.70517	0.70497	0.70476	0.70456	0.70435
0.70314	0.70294	0.70274	0.70254	0.70234
0.70116	0.70096	0.70076	0.70057	0.70037
0.69921	0.69902	0.69883	0.69863	0.69844
0.69731	0.69712	0.69693	0.69674	0.69656
0.69544	0.69526	0.69507	0.69489	0.69471
0.69362	0.69344	0.69326	0.69308	0.69290
0.69183	0.69165	0.69148	0.69130	0.69113
0.69008	0.68991	0.68973	0.68956	0.68939
0.68837	0.68820	0.68803	0.68786	0.68769
0.68669	0.68652	0.68636	0.68619	0.68603
0.68504	0.68488	0.68472	0.68456	0.68440
0.68343	0.68327	0.68311	0.68295	0.68279
0.68185	0.68169	0.68154	0.68138	0.68122
0.68029	0.68014	0.67999	0.67983	0.67968
0.67876	0.67861	0.67846	0.67831	0.67816
0.67726	0.67711	0.67696	0.67681	0.67666
0.67577	0.67562	0.67547	0.67532	0.67518
0.67429	0.67415	0.67400	0.67385	0.67371
0.67283	0.67268	0.67253	0.67239	0.67224
0.67136	0.67122	0.67107	0.67092	0.67078
0.66989	0.66975	0.66960	0.66945	0.66930
0.66841	0.66826	0.66811	0.66796	0.66781
0.66689	0.66674	0.66658	0.66643	0.66627
0.66533	0.66517	0.66501	0.66484	0.66468
0.66368	0.66351	0.66334	0.66316	0.66299
0.66190	0.66171	0.66151	0.66131	0.66111

TABLE 6.4 Madelung Constants for Several Structure Types

Structure	Madelung constant
Wurtzite (ZnS)	1.641
Sodium chloride (NaCl)	1.748
Cesium chloride (CsCl)	1.763
β-quartz (SiO$_2$)	2.220
Rutile (T$_1$O$_2$)	2.408
Fluorite (CaF$_2$)	2.519
Corundum (Al$_2$O$_3$)	4.172

The force holding the solid together is the derivative of the potential with respect to distance, therefore,

$$F = \left(\frac{dU}{dr}\right)_T = \frac{Z_c Z_a e^2}{r^2} - \frac{nb}{r^{(n+1)}}$$

and, at equilibrium, the attractive and repulsive forces must be zero, from which we find

$$b = \frac{Z_c Z_a e^2 r_0^{(n+1)}}{n}$$

and substituting this for b, in Eq. (6.13), at equilibrium,

$$U = -\frac{A Z_c Z_a e^2}{r_0}\left(1 - \frac{1}{n}\right) \tag{6.14}$$

where r_0 is the equilibrium separation distance between ions.

From thermodynamics we have Maxwell's equation, which relates the pressure to internal energy as

$$P = -\left(\frac{\partial U}{\partial V}\right)_T$$

and since the bulk modulus is

$$B = -V\left(\frac{\partial P}{\partial V}\right)_T$$

we see that we can also define it as

$$B = V\left(\frac{\partial^2 U}{\partial V^2}\right)_T \tag{6.15}$$

Therefore, taking the second derivative of V with respect to r, we have

$$\left(\frac{\partial^2 U}{\partial V^2}\right)_T = -\frac{2AZ_c Z_a e^2}{r_0^3}\left(1 - \frac{1}{n}\right) \qquad (6.16)$$

Now, the thermodynamic variables are P, V, and T, so it is necessary to change variables from r to V. To do this merely involves application of the chain rule to our differentiation, thus

$$\frac{dU}{dV} = \left(\frac{dU}{dr}\right)\left(\frac{dr}{dV}\right)$$

$$\frac{d^2U}{dV^2} = \left(\frac{dU}{dr}\right)\left(\frac{d^2r}{dV^2}\right) + \left(\frac{dr}{dV}\right)^2\left(\frac{d^2U}{dr^2}\right)$$

Since at equilibrium, $dU/dr = 0$, this reduces to

$$\frac{d^2U}{dV^2} = \left(\frac{dr}{dV}\right)^2\left(\frac{d^2U}{dr^2}\right) \qquad (6.17)$$

and dr/dV is simply

$$\frac{dr}{dV} = \frac{1}{3}\left(\frac{r}{V}\right) \qquad (6.18)$$

Substituting from Eqs. (6.18), (6.17), and (6.16) into (6.15), we obtain the desired relationship

$$B_0 V_0 = \frac{AZ_c Z_a e^2}{9r_0}(n-1) \qquad (6.19)$$

We define a parameter Ψ as

$$\Psi = \frac{B_0 V_0}{Z_c Z_a e^2} = \frac{A(n-1)}{9r_0} = \text{constant} \qquad (6.20)$$

and it is in this form that the bulk modulus–volume relationship is most useful. It is first necessary to demonstrate that Ψ is a constant. This has been done by Anderson and Anderson,[8] who found that for a large variety of compounds

$$\Psi = 0.157 \pm 0.005$$

Substituting this into Eq. (6.20), we have

$$\frac{B_0 V_0}{Z_c Z_a e^2} = 0.157 \pm 0.005 \qquad (6.21)$$

Taking logarithms of (6.21), we obtain

$$\ln B_0 = -\ln V_0 + [\ln Z_c Z_a + \ln(0.157 e^2)] \qquad (6.22)$$

where the last term in the brackets on the right is just a constant. Plotting $\ln B_0$ $\ln V_0$ for compounds with isomorphous electronic structures should yield straight lines with slopes of -1. Loci of different types of compounds would be separated by the intercept, which is the term in square brackets in (6.22), and this difference will be governed by the term $\ln(Z_c Z_a)$. For compounds containing more than one cation, the term $Z_c Z_a$ is calculated by

$$Z_c Z_a = \sum_{}^{p} \frac{\chi_i Z_i Z_a}{p}$$

where χ_i is the number of cations of valence Z_i, and p is the total number of cations in the chemical formula.

Two points should be kept in mind concerning the development of Eq. (6.21). First, the bulk modulus calculated in this way is the *isothermal* one. To compare it with the adiabatic bulk modulus, one must first convert it by the method discussed in Chap. 2. Second, the derivation is based upon the potential at absolute zero, since thermal effects have been ignored. Strictly speaking, therefore, it applies only at absolute zero. Fortunately, the contribution of thermal energy to the bulk modulus is small between absolute zero and room temperature for compounds with elevated melting temperatures, so that it yields values which are in reasonable agreement with room-temperature values of the isothermal bulk modulus.

6.3.2 Solid Solutions. In the preceding section we established a means of calculating the bulk modulus from the volume. Using Eq. (6.21), it is possible to calculate the bulk modulus of pure compounds. In many instances, it is desirable to determine what the bulk modulus of members of a solid solution would be. This is readily done with a direct thermodynamic approach.

We may express the volume for a solution of overall composition j as the sum of products formed by the molar volume of the pure constituents forming the solution and the mole fraction appropriate to each constituent, plus a volume change on mixing if the mixing is not ideal. This may be expressed as[9]

$$V_j = \sum_{i=1}^{n} V_i X_i + \Delta V_j \qquad (6.23)$$

where V_j = molar volume of solution of composition j
V_i = molar volume
X_i = mole fraction of the ith component of the solution

Differentiating Eq. (6.23) with respect to pressure and at constant temperature and composition,

$$\left(\frac{\partial V_j}{\partial P}\right)_{T,X} = \sum_{i=1}^{n} X_i \left(\frac{\partial V_i}{\partial P}\right)_{T,X} + \left(\frac{\partial (\Delta V_j)}{\partial P}\right)_{T,X}$$

and if ΔV_j is a function of *composition only*, then this reduces to

$$V_j \beta_j = \sum_{i=1}^{n} X_i V_i \beta_i \tag{6.24}$$

where β is the compressibility (reciprocal of the bulk modulus).

Thus, application of Eq. (6.21) together with (6.24) makes it possible to calculate the bulk modulus both of pure compounds and of solid solutions formed from them (see Fig. 6.2).

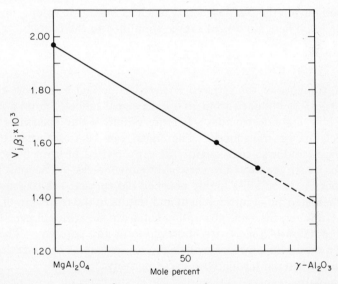

Fig. 6.2 The functon $\Sigma(X_i V_i \beta_i)$ as a function of molar composition. Circles denote measured data.

6.4 Use of the Infrared Reflection of a Diatomic Solid to Determine Bulk Modulus

This method is of limited usefulness since it applies only to diatomic solids. It has been used successfully to find the bulk modulus of MgO, SiC, and ZnO. It would presumably be useful for CoO, NiO, MnO, BN, and ZnS, but not for Fe_2O_3, SiO_2, etc. The method and its application to MgO have been described by Anderson and Glynn.[10]

The infrared reflection spectrum analysis can be used to determine the bulk modulus of diatomic solids. A general equation given by Szigeti[11]

$$B = \left(\frac{r_0}{3u}\right)\left(\frac{\varepsilon_0 + 2}{\varepsilon_\infty + 2}\right)\bar{m}\omega_0^2 \qquad (6.25)$$

where r_0 = interatomic spacing
u = volume of ion pair
ε_0 = dielectric constant
ε_∞ = square of refractive index n
$\bar{m}$ = reduced mass
ω_0 = reststrahlen frequency

Since the absorption wavelength (λ_0) is $2\pi C/\omega_0$, Eq. (6.25) becomes

$$B\lambda_0^2 = 4\pi^2 C^2 \frac{r_0^2 \bar{m}}{3u} \frac{\varepsilon_0 + 2}{\varepsilon_\infty + 2} \qquad (6.26)$$

For a solid having the NaCl structure, such as MgO, the volume of the ion pair is $2r_0^3$. Then, Eq. (6.26) can be simplified to the form

$$B\lambda_0^2 = 4\pi^2 C^2 \frac{\bar{m}(\varepsilon_0 + 2)}{6r_0(\varepsilon_\infty + 2)} \qquad (6.27)$$

The advantages of the infrared reflectivity method stem from the fact that the signal is obtained from an optical signal reflected from a surface. The method is nondestructive. It requires only a thin specimen and no holder, clamp, or transducer. The signal can be transmitted through appropriate windows; consequently, it can be used on thin films, or on specimens which are heated to very high temperatures. The most critical requirement is to obtain a highly polished flat surface. A coarse surface causes the scattering of infrared light and results in the reduction of reflectivity. However, the same absorption band can be achieved even though such a reduction of reflectivity occurs, due to any porosity. Therefore, the bulk modulus of a material with zero porosity can be obtained from the data using a material having fairly large porosity.

The chief disadvantage is that only one elastic constant is found, so that the shear constant or Poisson's ratio cannot be determined. Another disadvantage is that the value of bulk modulus is only determinable to within a few percent. However, it may be the only method available to examine the elasticity of a thin film. It is unfortunate that the more complicated compounds show structured reflection bands which cannot be analyzed, so that the applicability of the method is limited to diatomic solids at the present time. A description of the results found by Anderson and Glynn[10] for MgO is given below.

The reflection band was measured on a polycrystalline MgO specimen. The reflectivities were determined using a point-by-point measurement. The intensity of the reflection from the sample onto a thermocouple was compared with that of the reflection from the front surface of a high-quality rhodium mirror. The mirror was assumed to have a reflectivity of 97 percent for wavelengths greater than 1 μm.

By analysis of the reflection band between 4 μm and 30 μm, the absorption wavelength λ_0 was found to be 25.0 ± 0.1 μm. The index of refraction was 1.783, and the dielectric constant was 9.8. These measurements used in Eq. (6.27) gave the value of the bulk modulus as $B = 1,640$ kbar, which is in good agreement with the value determined from acoustic data. The main source of error is in the determination of λ_0 and the dielectric constant.

6.5 Estimating the Temperature Derivatives of Elastic Moduli

Wachtman[12] reported an empirical relationship for Young's modulus as a function of temperature. The equation has the form

$$E = E_0 - b_1 T e^{(-T_0/T)} \qquad (6.28)$$

where E_0 = value of Young's modulus at absolute zero
b and T_0 = experimentally determined constants

The equation was tested and found to hold very well for Al_2O_3 and ThO_2. However, as the authors of the report noted at the time, "No theoretical justification (of the equation) is known, ... and the physical interpretation of the parameters b and T_0 presents a challenging theoretical problem."

From this starting point, the problem was approached from the view of determining a theoretical basis for Eq. (6.28). Anderson,[13] who had been exploring the relationships controlling the bulk modulus, found that an analogous equation also held:

$$B = B_0 - bTe^{(-T_0/T)} \qquad (6.29)$$

This is a more fundamental relationship than (6.28) because here the bulk modulus is a thermodynamic property, whereas Young's modulus is not.

The parameter b was found to be related to fundamental constants by the equation

$$b = \frac{3R\gamma\delta}{V_0} \qquad (6.30)$$

where R = gas constant
γ = first Grüneisen parameter, which is the well-known parameter $\alpha B_s/C_p\rho$
V_0 = mean atomic volume $M/p\rho$
δ = second Grüneisen parameter given by Eq. (6.31)

$$\delta = -\frac{d(\ln B_s)/dT}{d(\ln V)/dT} = -\frac{1}{\alpha B_s}\left(\frac{dB_s}{dT}\right) \qquad (6.31)$$

Both γ and δ have been shown to be independent of temperature at high temperatures and hence may be regarded as constants. The parameter T_0 plays the role of a characteristic temperature, related to the Debye temperature θ, where the quantity $3RTe^{(-T_0/T)}$ is an empirical representation of the energy $\int C_v dT$ over the experimental range. To derive Eq. (6.29), we relate the two Grüneisen parameters in the following way:

$$\alpha = \frac{-1}{\delta B_s}\left(\frac{dB_s}{dT}\right) = \frac{\gamma C_p}{V B_s}$$

which becomes

$$\left(\frac{dB_s}{dT}\right) = -\gamma\delta\left(\frac{C_p}{V}\right) \tag{6.32}$$

which then becomes

$$\int dB_s = -\gamma\delta \int \frac{C_p}{V} dT \tag{6.33}$$

and employing the approximation,

$$\int_0^T \frac{C_p}{V} dT \cong \frac{1}{V_0}\int_0^T C_v \, dT$$

Since $\int_0^T C_v dT$ is the thermal energy of the solid $\mathscr{E}$, we have, upon integration,

$$B_s = B_0 - \left(\frac{\gamma\delta}{V_0}\right)\mathscr{E} \tag{6.34}$$

Using the Debye approximation, the thermal energy per gram atom is

$$\mathscr{E} = 3RTf\left(\frac{\theta}{T}\right) \tag{6.35}$$

where θ is the Debye temperature of the solid and the function $f(\theta/T)$ is of the form

$$f\left(\frac{\theta}{T}\right) \simeq \frac{x}{e^x - 1}$$

Combining Eqs. (6.34) and (6.35) yields

$$B_s = B_0 - \frac{3R\gamma\delta T}{V_0}f\left(\frac{\theta}{T}\right) \tag{6.36}$$

from which we find that the parameter b is given by $3R\gamma\delta/V_0$.

If we expand $f(\theta/T)$, we obtain

$$f\left(\frac{\theta}{T}\right) \simeq \left[1 + \frac{\theta}{2T} + \frac{2}{3}\left(\frac{\theta}{2T}\right)^2 + \cdots\right]^{-1}$$

and similarly, if we expand the exponential term in Eq. (6.29), we have

$$e^{-(T_0/T)} \simeq \left[1 + \frac{T_0}{T} + \frac{1}{2}\left(\frac{T_0}{T}\right)^2 + \cdots\right]^{-1}$$

At high temperatures both of these series converge to the same value, and we note that

$$T_0 \simeq \frac{\theta}{2}$$

which is the relation between T_0 in Eq. (6.29) and the characteristic Debye temperature.

The reason that Wachtman found his original equation, involving Young's modulus, to be a successful one is also interesting. To see why this was so, it is useful to relate Young's modulus to the bulk modulus and Poisson's ratio:

$$E = 3B_s(1 - 2\sigma)$$

and differentiating this equation with respect to temperature,

$$\frac{dE}{dT} = 3(1 - 2\sigma)\left(\frac{dB_s}{dT}\right) - 6B_s\left(\frac{d\sigma}{dT}\right) \tag{6.37}$$

For those materials for which $d\sigma/dT$ is a constant (6.36) may be integrated to yield

$$E - E_0 = 3(1 - 2\sigma)\int\left(\frac{dB_s}{dT}\right)dT - 6B_s\frac{d\sigma}{dT} \tag{6.38}$$

from which we deduce that, for any material for which $d\sigma/dT$ is small and independent of temperature, the relation found by Wachtman will apply as a special case.

Since we can also relate the shear modulus to the bulk modulus, an analogous equation can also be obtained. Explicitly, we have

$$G = 3B\frac{1 - 2\sigma}{1 + 2\sigma} \tag{6.39}$$

and after differentiating this equation, we have

$$\frac{dG}{dT} = \frac{3(1 - 2\sigma)}{(1 + 2\sigma)}\left(\frac{dB_s}{dT}\right) - \frac{12B}{(1 + 2\sigma)^2}\left(\frac{d\sigma}{dT}\right)$$

and for $d\sigma/dT$ small and constant,

$$G - G_0 = \frac{3(1 - 2\sigma)}{(1 + 2\sigma)}\int\left(\frac{dB}{dT}\right)dT - \frac{12B}{(1 + 2\sigma)^2}\left(\frac{d\sigma}{dT}\right) \tag{6.40}$$

Since $d\sigma/dT$ is of the order of 10^{-5} for many compounds, particularly oxides and silicates, the last term may safely be ignored in the calculation, so that Eqs. (6.38) and (6.40) become

$$E = E_0 + 3(1 - 2\sigma) \int \left(\frac{dB_s}{dT}\right) dT \qquad (6.41a)$$

$$G = G_0 + \frac{3(1 - 2\sigma)}{(1 + 2\sigma)} \int \left(\frac{dB_s}{dT}\right) dT \qquad (6.41b)$$

There are two convenient ways to utilize Eqs. (6.29) and (6.41a) and (6.41b) (remembering the restriction on the applicability of the latter two). The problem centers about the function $f(\theta/T)$. One approach is through the functional relationship between the internal energy and the Debye temperature. If the Debye temperature of the solid is known, then values of $\mathscr{E}/T$ may be obtained from tables relating $\mathscr{E}/T$ to θ/T. In this case, Eq. (6.36) is used in the form

$$B_s = B_0 - \frac{\gamma\delta}{V_0} T \frac{\mathscr{E}(\theta/T)}{T} \qquad (6.42)$$

The value of γ may be calculated from available data on thermal expansivity, heat capacity, and bulk modulus. The value of δ has been found to be about 4 for oxides and silicates. It may also be found for ionic solids from

$$\delta = \left(\frac{m+n}{3} + 2 - \gamma\right) \qquad (6.43)$$

where m and n are the attractive and repulsive exponents of the Born potential. For many alkali halides, $m = 1$ and $n = 9$.

A second useful formulation for the evaluation of Eq. (6.29) is based on the availability of thermal expansivity and heat-capacity data. We return to Eq. (6.33):

$$B_s - B_0 = -\gamma\delta \int_{298}^{T} \frac{C_p}{V} dT$$

(where the reference temperature is now taken to be 298°K). We note that the integrand may be expressed as

$$\frac{1}{V_{298}} \int_{298}^{T} C_p(1 + \alpha T) \, dT$$

since at high temperature we may safely ignore higher-order terms in thermal expansivity. Integrating we then have

$$\frac{1}{V_{298}} \left(\Delta H - \int_{298}^{T} C_p \alpha T dT\right)$$

where ΔH is the enthalpy difference between 298 and T. The remaining integrand is approximated quite well by $\alpha T\Delta H/2$, since at high temperatures α approaches a constant value and $\int C_p dT$ is linear. We may then formulate the alternative expression for evaluating (6.36) based upon the availability of enthalpy and thermal expansivity data:

$$B_s = B_0 - \frac{\gamma \delta \Delta H}{V_{298}}\left(1 - \frac{\alpha T}{2}\right) \tag{6.44}$$

Since the compressional and shear velocities are related to the bulk modulus, and shear velocity to the shear modulus, Eqs. (6.41a) and (6.41b) can be combined to estimate these velocities as a function of temperature.

6.6 Estimating the Pressure Derivative of the Bulk Modulus

6.6.1 Estimation from the Grüneisen Parameter.
The relationship between the pressure derivative of the bulk modulus $(\partial B/\partial P)_T$ and the Grüneisen constant of a solid arises from considerations of the dependence of vibrational frequencies with volume. Assumptions have to be made to make the theory tractable, and, consequently, there is some arbitrariness in the actual functional relationship. The most widely used is the Dugdale-MacDonald relationship,[14]

$$\left(\frac{\partial B_0}{\partial P}\right)_T = 2\gamma + 1 \tag{6.45}$$

Here γ has the usual definition [see Eq. (6.30)]. The value of γ can be found from a knowledge of thermal expansivity, the ambient value of the bulk modulus, the ambient density, and the values the of specific heat. Equation (6.42) was used by Anderson[15] to estimate the compressed volume at high pressures. Rather good results were achieved for lead, tungsten, and vanadium, even up to pressures of 1 mbar. This equation yields an approximate value, as is demonstrated in Table 6.5. In general it tends to underestimate the values that have been determined from direct measurement and serves at best as a rough approximation.

6.6.2 Estimation of dB/dP from the Repulsion Exponent in the Born Potential.
In Sec. 6.3.1. we developed equations for determining the bulk modulus from the volume based upon the Born formulation of the interionic potential. In this section, we will extend the treatment to develop an equation relating the pressure derivative of the bulk modulus to the exponent in the repulsion term of the potential. In measuring the pressure derivative of the bulk modulus, we change the interionic spacing from its equilibrium position r_0. Hence the formulation used earlier,

TABLE 6.5 Estimating dB_0/dP from the Grüneisen Parameter

		dB_0/dP	
Compound	γ	Calculated	Measured
$\alpha - A_2lO_3$	1.32	3.6	4.0
MgO	1.54	4.1	3.9
CaO	1.19	3.4	5.2
ZnO	0.81	2.6	4.8
BeO	1.27	3.5	4.4
$MgAl_2O_4$	1.13	3.3	4.3
$NiFe_2O_4$	1.24	3.4	4.4
$\alpha - Fe_2O_3$	1.99	5.0	4.5
Mg_2SiO_4	1.17	3.3	5.4
$\alpha - SiO_2$	0.70	2.4	6.4

which applies only to the equilibrium situation, must be generalized, and for this reason we must keep r as a variable in our equations. We start with the Born potential (6.13), but now express the parameter b, at the equilibrium condition, as

$$\frac{b}{r_0^n} = \frac{AZ_c Z_a e^2}{r_0}$$

and substituting into the equation for the potential,

$$U = \frac{AZ_c Z_a e^2}{r_0} \left[\left(\frac{r_0}{r}\right) - \frac{1}{n}\left(\frac{r_0}{r}\right)^n \right] \quad (6.46)$$

Note: Equation (6.46) is identical, in principle, to (6.14) except that now we keep r as an explicit variable. Differentiating with respect to r and multiplying the result by r yields

$$r\left(\frac{dU}{dr}\right) = \frac{AZ_c Z_a e^2}{r_0} \left[\left(\frac{r_0}{r}\right)^n - \left(\frac{r_0}{r}\right) \right]$$

Now the pressure is related to the derivative of the potential through

$$P = -\frac{dU}{dV}$$

or

$$P = \left(\frac{dU}{dr}\right)\left(\frac{dr}{dV}\right) = -\frac{1}{3V}\left(\frac{rdU}{dr}\right)$$

therefore we now have

$$P = \frac{1}{3V} \frac{AZ_c Z_a e^2}{r_0} \left[\left(\frac{r_0}{r}\right)^n - \left(\frac{r_0}{r}\right) \right]$$

It is now convenient to introduce the equilibrium volume V_0, which is related to V by

$$\frac{1}{V} = \frac{1}{V_0}\left(\frac{r_0}{r}\right)^3$$

hence we have

$$P = \frac{AZ_c Z_a e^2}{3V_0 r_0}\left[\left(\frac{\rho}{\rho_0}\right)^{(n+3)/3} + \left(\frac{\rho}{\rho_0}\right)^{4/3}\right] \quad (6.47)$$

where we have introduced the density in place of r in the term inside the square brackets. Rearranging Eq. (6.19),

$$\frac{3B_0}{(n-1)} = \frac{AZ_c Z_a e^2}{3V_0 r_0}$$

and substituting into Eq. (6.46),

$$P = \frac{3B_0}{(n-1)}\left[\left(\frac{\rho}{\rho_0}\right)^{(n+3)/3} - \left(\frac{\rho}{\rho_0}\right)^{4/3}\right] \quad (6.48)$$

Differentiating this equation with respect to density and multiplying by the density yields a general equation for the bulk modulus [since $B = 1/V(dP/dV) = \rho(dP/d\rho)$] based upon the Born potential:

$$B = \frac{B_0}{(n-1)}\left[(n+3)\left(\frac{\rho}{\rho_0}\right)^{(n+3)/3} - 4\left(\frac{\rho}{\rho_0}\right)^{4/3}\right] \quad (6.49)$$

To obtain dB/dP from (6.49), differentiate it with respect to the density [noting that $dB/dP = \rho/B(dB/d\rho)$], and again multiply through by the density to obtain

$$\frac{dB}{dP} = \frac{1}{3}\left[\frac{(n+3)^2\left(\frac{\rho}{\rho_0}\right)^{(n+3)/3} - 16\left(\frac{\rho}{\rho_0}\right)^{4/3}}{(n+3)\left(\frac{\rho}{\rho_0}\right)^{(n+3)/3} - 4\left(\frac{\rho}{\rho_0}\right)^{4/3}}\right]$$

This can be simplified by first factoring out $(\rho/\rho_0)^{4/3}$ and then adding $(4n - 4n)$ to the numerator to yield

$$\frac{dB}{dP} = \frac{n+3}{3} + \frac{4}{3}\left[\frac{(n-1)}{(n+3)(\rho/\rho_0)^{(n-1)/3} - 4}\right] \quad (6.50)$$

Then at zero pressure, $\rho = \rho_0$ so that (6.50) reduces to

$$\frac{dB_0}{dp} = \frac{n+7}{3} \quad (6.51)$$

and, at extremely high pressure, (ρ/ρ_0) is very large and the term in square brackets approaches zero, in which case (6.50) becomes

$$\frac{dB_\infty}{dP} = \frac{n+3}{3} \qquad (6.52)$$

Equation (6.51) refers to the initial slope of the bulk modulus–pressure curve, and corresponds to the value measured at low pressure. Equation (6.52) applies at very high pressures of the sort achieved in shock wave experiments. Equation (6.51) places some severe restrictions on the possible range of values of dB_0/dP. Since the value of n ranges from about 4 to 12, values of dB_0/dP should be expected to lie within the range 3.7 to 6.3. Values of dB_0/dP very much outside this range should be considered suspect. For example, a reported value of 3 for dB_0/dP would imply a repulsion exponent of only 2, which would result in a repulsion not very much greater than the attractive potential—and hence indicate a very highly compressible solid.

6.6.3 Determination of Bulk Modulus B and dB/dP from High-Pressure P-V Data.

If one has isothermal data showing how the density ratio varies with pressure, one can infer from such data the value of the bulk modulus B, and the pressure derivative dB_0/dP at that temperature. There are several formulas by which one can find these parameters with curve-fitting techniques. One common relation is the Murnaghan equation (where $dB_0/dP = B_0'$):[16]

$$\frac{P}{B_0} = \frac{1}{B_0'}\left[\left(\frac{\rho}{\rho_0}\right)^{B_0'} - 1\right] \qquad (6.53)$$

Another is the Birch equation:[16]

$$\frac{P}{B_0} = \frac{3}{2}\left[\left(\frac{\rho}{\rho_0}\right)^{7/3} - \left(\frac{\rho}{\rho_0}\right)^{5/3}\right]\left\{1 - \frac{3}{4}(4 - B_0')\left[\left(\frac{\rho}{\rho_0}\right)^{2/3} - 1\right]\right\} \qquad (6.54)$$

Many pairs of values of B_0 and B_0' can satisfy the same P-V data. Therefore if static compression data alone are used to determine both B_0 and B_0', a large error may exist in either or both values. If, however, one of these values is known independently of the static compression data, the other value can be determined by this method with considerably less uncertainty.

REFERENCES

1. Anderson, O. L.: An Approximate Method of Measuring the Specific Heat, *J. Geophys. Res. Lett.*, **70**:4726 (1965).
2. Alers, G. A.: Use of Sound Velocity Measurements in Determining the Debye Temperature of Solids, in W. P. Mason (ed.), "Physical Acoustics," vol. 3B, chap. 1, Academic, New York, 1965.

3. Barron, T. H. K., W. T. Berg, and J. A. Morrison: Properties of Alkali Halide Crystals at Low Temperatures, *Phys. Rev.*, **115**:1439 (1959).
4. Anderson, O. L.: The Debye Temperature of Vitreous Silica, *J. Phys. Chem. Sol.*, **12**:41 (1959).
5. Anderson, O. L., and N. Soga: A Simplified Method for Calculating the Elastic Moduli of Ceramic Powder from Compressibility and Debye Temperature, *J. Am. Ceram. Soc.*, **49**:318 (1966).
6. Weir, C. E.: Isothermal Compressibilities of Alkaline Earth Oxides at 21°C, *J. Res. Natl. Bur. Stand.*, **56**:187 (1956).
7. Potter, R. F.: The Ionic Character and Elastic Moduli of Zincblende Lattices, *J. Phys. Chem. Sol.*, **3**:223 (1957).
8. Anderson, D. L., and O. L. Anderson: The Bulk Modulus–Volume Relationship for Oxides, *J. Geophys. Res.*, **75**:3494 (1970).
9. Schreiber, E.: The Effect of Solid Solution upon the Bulk Modulus and Its Pressure Derivative: Implications for Equations of State, *Earth Planet. Sci. Lett.*, **7**, 137 (1969).
10. Anderson, O. L., and P. Glynn: Measurement of Compressibility in Polycrystalline MgO Using the Reflectivity Method, *J. Phys. Chem. Sol.*, **26**:1961 (1965).
11. Szigeti, B.: Compressibility and Absorption Frequency of Ionic Crystals, *Proc. Roy. Soc. Lond.*, **204**:51 (1950).
12. Wachtman, Jr., J. B., W. E. Tefft, D. G. Lam, Jr., and C. S. Apstein: Exponential Temperature Dependence of Young's Modulus for Several Oxides, *Phys. Rev.*, **122**:1754 (1961).
13. Anderson, O. L.: A Derivation of Wachtman's Equation for the Temperature Dependence of Elastic Moduli of Oxide Compounds, *Phys. Rev.*, **144**:553 (1966).
14. Dugdale, J. S., and D. K. C. MacDonald: Thermal Expansion of Solids, *Phys. Rev.*, **89**:832 (1953).
15. Anderson, O. L.: Use of Ultrasonic Measurements at Modest Pressure to Estimate High Pressure Compression, *J. Phys. Chem. Sol.*, **26**:547 (1966).
16. Murnaghan, F. D.: Compressibility of Media under Extreme Pressure, *Proc. Natl. Acad. Sci. U.S.*, **50**: 244 (1944).
17. Birch, F.: The Effect of Pressure upon the Elastic Constants of Isotropic Solids according to Murnaghan's Theory of Finite Strain, *J. Appl. Phys.*, **9**:279 (1938).

CHAPTER SEVEN

The Pressure and Temperature Derivatives of Elastic Constants and Thermodynamic Functions

7.1 Introduction

We review, in this chapter, a number of treatments concerning the pressure derivatives of elastic constants. We are concerned mainly with the transformations from the raw data to certain elastic constants and their derivatives, and the transformation of higher-order elastic constants from one mode of expression to another. The object is to enumerate the paths leading from elastic-constant data as these are found in the literature or measured in the laboratory to parameters most useful in thermodynamics.

Actual data on elastic constants are listed here only when needed to illustrate a method of computation. The reader is referred to Ref. 14 for a compilation of data on such constants before and after they are transformed.

The connection between elastic constants and ordinary thermodynamics —pressure, volume, temperature space, i.e., equations of state $P = f(V, T)$— requires only a single scalar invariant of elastic constants, namely the bulk modulus. On the other hand, elastic constants and their pressure and temperature derivatives are fourth-rank tensors. For this reason, the bulk modulus and its derivatives are of primary importance as the connection between acoustics and thermodynamics. Information on how the volume varies with pressure and temperature is generally available, but information on how Poisson's ratio or the shear modulus varies with pressure or temperature is not generally available from most direct experimental data. Therefore, the main emphasis is concerned with various higher derivatives of combinations of V, P, and T. The only combination of elastic constants needed, therefore, is that which yields the bulk modulus, since it directly relates pressure and volume. In the general sense,

$$B_x = -V\left(\frac{\partial P}{\partial V}\right)_x \tag{7.1}$$

where x represents the conditions of the experiment: if adiabatic x is S, isothermal x is T. Both adiabatic and isothermal values of the bulk modulus arise because dynamic determination of elastic constants is taken under adiabatic conditions, but direct measurements of volume change due to a pressure change are performed isothermally. The zero pressure value of the bulk modulus at room temperature is most commonly available. In our nomenclature, when it is necessary to specify the temperature and pressure at which B is measured, we write, for example,

$$B_x(0, T_0) = -V\left(\frac{\partial P}{\partial V}\right)_x \qquad P \to 0,\ T = T_0 \tag{7.2}$$

The higher derivatives of the bulk modulus are important. We indicate pressure differentiation by a prime when the derivative is taken as P approaches zero.

Unless otherwise indicated, the prime partial also means that the derivatives with respect to pressure are taken at constant temperature. For example, one differentiation of the adiabatic bulk modulus with respect to P at *constant temperature* and at zero pressure could be written

$$B'_S = \left\{\frac{\partial[-V(\partial P/\partial V)_S]}{\partial P}\right\}_{T_0 = T,\ P \to 0} \tag{7.3}$$

In case the derivative is taken at higher pressure, or if we wish to specify the temperature, we write out the long notation $B'_S(P, T)$.

We now emphasize the point that not all elastic constants of a solid are pertinent to thermodynamics. In a cubic solid the three second-order independent elastic constants are c_{11}, c_{12}, and c_{44}. In our nomenclature, we write for the bulk modulus

$$B_S = \frac{(c_{11} + 2c_{12})}{3} \tag{7.4}$$

so that c_{44} does not appear in the bulk modulus or its pressure derivative. [Isothermal single-crystal constants are indicated by roman (not italic) type, e.g., c_{11}, c_{12}, c_{44}].

A similar situation occurs for the higher derivatives with respect to pressure. For the two least symmetrical classes of cubic crystals, there are eight independent *third-order* elastic constants; these are C_{111}, C_{112}, C_{113}, C_{114}, C_{144}, C_{155}, C_{166}, C_{456}. Of these, only four appear in the isothermal-pressure derivative of the bulk modulus.[1] We have

$$B'_S = -\frac{C_{111} + 5C_{112} + C_{113} + 2C_{123}}{3(c_{11} + c_{12})} - \frac{\alpha \gamma T}{3} \qquad (7.5)$$

where α is the volumetric coefficient of the thermal expansion, and γ is the Grüneisen constant given in our notation as

$$\gamma = \frac{\alpha B_S}{\rho C_P} = \frac{\alpha B_T}{\rho C_V} \qquad (7.6)$$

where ρ = density
C_P and C_V = specific heat at constant pressure and volume, respectively

The $C_{\lambda\mu\nu}$ in Eq. (7.5) represents the third-order constants as defined by Thurston and Brugger.[2]

For the most symmetrical classes of cubic crystals, there are only six independent third-order constants, because, in these crystals, $C_{113} = C_{112}$ and $C_{166} = C_{155}$. In this case, Eq. (7.5) undergoes obvious simplification.

If one considers the pressure derivative of the bulk modulus in terms of the second-order elastic constants themselves, for example, dc_{11}/dP, dc_{44}/dP, and dc_{12}/dP, etc., as the data are sometimes reported, then for cubic crystals

$$B'_S = \frac{1}{3}\left[\left(\frac{\partial c_{11}}{\partial P}\right)_T + 2\left(\frac{\partial c_{12}}{\partial P}\right)_T\right] - \frac{\alpha \gamma_T}{3} \qquad (7.7)$$

Again it can be seen that not all the pressure derivatives that are measurable enter into thermodynamic functions.

7.2 Pressure Dependence of Elastic Constants of Cubic Crystals

The proper use of information on the changes of elastic constants with pressure requires that careful attention be given to definitions. How is an elastic constant to be defined as a function of pressure?

One attractive definition relates elastic constants to the values of ρv^2 (ρ = density, v = propagation velocity) for the various modes of propagation in a crystal, where the same formulas are used with all pressures. For example, in a cubic crystal, $c_{11}(P)$ is the value of ρv^2 for a longitudinal mode propagating along a crystal axis. To calculate it from a measured transit time, one needs the density and path length at the pressure P. This definition, historically, was used in the ultrasonics literature prior to recent interest in third-order elastic constants. We follow Thurston[3,4] in calling these functions of pressure the *effective* elastic constants. Thurston's work shows that effective elastic constants are not the only means to define the constants, and, indeed, in some contexts other definitions are superior.

It turns out that effective elastic constants are not as tractable in equations involving third-order elastic constants as in other definitions. Elastic constants of the nth order are conveniently defined as the nth derivatives of the energy per unit volume with respect to strain components, the strain being measured from the configuration of zero pressure. (The adiabatic constants are obtained by differentiating the internal energy at constant entropy, and the isothermal ones by differentiating at constant temperature.) Elastic constants defined in this way are called *thermodynamic*. They may be obtained as a function of pressure simply by evaluating the pressure derivatives at the pressurized state.

At zero pressure, the effective and thermodynamic *constants* are the same, although they represent different functions of pressure, and they have different pressure derivatives even at zero pressure. The relations between the pressure derivatives of *thermodynamic* and *effective* elastic constants have been worked out by Thurston.[3,4] An appreciation of the distinction between pressure derivatives of effective and thermodynamic elastic constants is essential in relating pressure to measured third-order elastic constants. As far as equations of state are concerned, this distinction is important because the value of B'_x may have to be determined from data published either as third-order constants or as pressure derivatives of effective elastic constants.

From their definition, third-order constants are strain derivatives of second-order *thermodynamic* constants. Hence, by the chain rule differentiation formula, the pressure derivatives of the thermodynamic constants $\bar{c}_{\lambda\mu}$ may be expressed as

$$\frac{dc_{\lambda\mu}}{dP} = \sum_{\nu}^{6} C_{\lambda\mu\nu} \frac{dn_\nu}{dP} = -\sum_{\nu}^{6} C_{\lambda\mu\nu}(s_{\nu1} + s_{\nu2} + s_{\nu3}) \tag{7.8}$$

Here n is the appropriate strain component, and $s_{\nu 1}$ is the appropriate element of the compliance tensor. Further, the 6×6 compliance matrix is found by inverting the 6×6 second-order elastic constant matrix.

It may be noted that the pressure derivatives of the elastic constants have dimensionless units, whereas the third-order constants have the units of pressure—the same as the second-order constants—and the compliance $s_{\nu 1}$ has the dimensions of reciprocal pressure.

TABLE 7.1 Difference of Pressure Derivatives of Effective and Thermodynamic Elastic Coefficients $(c_{\lambda\nu} - \bar{c}_{\lambda\nu})$, Excluding Monoclinic and Triclinic Classes (after Thurston[3])

$s_i = s_{i1} + s_{i2} + s_{i3} \quad i = 1, 2, 3$

$-1 + (s_2 + s_3 - 3s_1)c_{11}$	$1 + (s_3 - s_1 - s_2)c_{12}$	$1 + (s_2 - s_1 - s_3)c_{13}$	$-s_1 c_{14}$	$(s_2 - 2s_1)c_{15}$	$(s_3 - 2s_1)c_{16}$
	$-1 + (s_1 + s_3 - 3s_2)c_{22}$	$1 + (s_1 - s_2 - s_3)c_{23}$	$(s_1 - 2s_2)c_{24}$	$-s_2 c_{25}$	$(s_3 - 2s_2)c_{26}$
		$-1 + (s_1 + s_2 - 3s_3)c_{33}$	0	0	0
			$-1 + (s_1 - s_2 - s_3)c_{44}$	$-s_3 c_{45}$	$-s_2 c_{46}$
				$-1 + (s_2 - s_1 - s_3)c_{55}$	$-s_1 c_{56}$
					$-1 + (s_3 - s_1 - s_2)c_{66}$

We note that since $d\bar{c}_{\lambda\mu}/dP$ is ordinarily evaluated at zero pressure, the values of s_{vi} are zero-pressure compliances; and since no distinction between thermodynamic and effective compliances need be made to calculate the pressure derivatives of the *effective* constants from the third-order constants, the difference tabulated by Thurston must be added to Eq. (7.8). These are shown in Table 7.1.

In processing ultrasonic data taken as a function of hydrostatic pressure, it has been traditional to calculate the effective elastic constants at the various pressures used. This requires values of density and path length at elevated pressure. Experience has shown that the effective elastic constants are very nearly linear functions of pressure in the measured range, so essentially all the obtainable information is contained in their initial pressure derivatives. As has been emphasized by Thurston,[1] these derivatives can be calculated directly from the intercepts and initial slopes of the measured frequency (or transit time) versus pressure curves without the values of density or path length at any elevated pressure. This improves the data-processing slightly.

The typical measurement technique consists of a measurement of frequency f (reciprocal transit time) at a given pressure P. Experimentalists following the tradition of McSkimin[5] exert great care in the measurement of pressure in order to find out if f is linear with pressure, and it is found that with the accuracy of the best measurements of pressure (1 part in 10^4), f is linear with pressure for many solids. (The measurement of f itself is sensitive to changes of a few parts in 10^5 using ultrasonic pulse superposition techniques.) The measurement of f is ordinarily not carried out above 10 kbar, because the errors in the measurement of pressure become too large to make the derivative $\Delta f/\Delta P$ very meaningful. For this reason, many experimental results are reported below 4 kbar, or even 2 kbar, depending upon the range of the pressure-measuring device available to the experimenter.

A mode distinguishes a direction of propagation in the crystal and the direction of the particle motion (transverse or longitudinal). For a cubic crystal, propagation of sound by eight modes gives enough information to calculate the elastic constants with five cross-checks. A sample showing the data for eight modes of MgO and the various cross-checks on the computation of *second-order* constants is given in Table 7.2.

Whenever the frequency f is found to be linear with pressure, the experiment can be summarized by a table showing the intercept at zero pressure f_0 and the slope $\Delta f/\Delta P$. The essential information is found in the slope divided by the intercept: $\Delta f/f_0 \Delta P$. This information for the first five modes measured at two different temperatures for MgO is shown in Table 7.3.

TABLE 7.2 Basic Data for Wave Velocities at 23°C (Excess Pressure = 0)

Designation	Mode	Direction of wave propagation	Direction of particle motion	Path length,* cm	Frequency, Hz	Velocity, km/s
v_1	Long	001	001	2.5787	352,730	9.09585
v_2	Shear	001	$1\bar{1}0$	2.5787	255,765	6.59539
v_3	Long	110	110	2.5269	392,048	9.90669
v_4	Shear	110	001	2.5269	261,039	6.59619
v_5	Shear	110	$1\bar{1}0$	2.5269	209,787	5.30111
v_6	Long	$1\bar{1}0$	$1\bar{1}0$	2.6203	378,102	9.90740
v_7	Shear	$1\bar{1}0$	001	2.6203	202,317	6.59608
v_8	Shear	$1\bar{1}0$	110	2.6203	251,730	5.30131

* Equals twice the specimen dimension.

A digression at this point is made to discuss the meaning of the measured linearity of f with P. A glance at the last column of Table 7.3 shows that the slope is very small. Experimentally, one cannot distinguish from the data between f linear and f^2 linear, but one of these quantities *must* be nonlinear with pressure. However, the assumption of the linearity of bulk modulus with pressure at very high pressure has been tested and appears to hold. From dimensional arguments, therefore, it seems that the f^2 linear law is more nearly correct and the f linear law is only an

TABLE 7.3 Best-Fit Straight Lines for Variations of f with Pressure

Mode	Slope, Hz/(lb/in²)	Slope, Hz/kbar	Intercept, Hz	Correction to f at 30,000 lb/in² 2.0681 kbar	$\Delta f/f_0 \Delta P$ per kbar
			$T = 23°C$		
1	0.017801	258.18	176,391 ($N = 2$)	-1.5	0.00146370
2	0.002399	34.795	127,888 ($N = 2$)	$+1.7$	0.00027203
3	0.011105	161.066	196,025 ($N = 2$)	-1.5	0.00082166
4	0.003637	52.75105	261,033 ($N = 1$)	$+6.5$	0.00040418
5	0.012202	176.977	104,917 ($N = 2$)	$+0.8$	0.00168682
			$T = -195.8°C$		
1	0.017848	258.87	179,095 ($N = 2$)	-3.0	0.00144544
2	0.002853	41.380	128,496 ($N = 2$)	$+2.2$	0.00032204
3	0.010623	154.076	197,504 ($N = 2$)	-1.7	0.00078012
4	0.004794	69.532	262,223 ($N = 1$)	$+6.5$	0.00053034
5	0.011604	168.304	107,637 ($N = 2$)	$+4.5$	0.00156362

apparent fit. It is a very good fit, however, and is quite sufficient for the calculation with a high accuracy of higher-order constants at zero pressure.

Using measured data, such as shown in Tables 7.2 and 7.3, the pressure derivatives of the elastic constants can be determined in two ways. The usual way is to calculate the effective elastic coefficients at the highest and lowest pressure and get the derivative by dividing the difference by the pressure. Geometrically, this is the chord of a function.

The first step is to determine the change in length due to pressure at the particular temperature of the measurement. This change of length depends upon the compressibility which, in turn, depends upon the elastic constants themselves. The calculation can be carried out using the method described by Cook[6] and outlined in Chap. 3. Knowledge of the values $f'/f_0 = \Delta f/f_0 \Delta P$ for modes v_1 and v_5, in Table 7.3, are sufficient to extract this information. The formula for length change is given in Eqs. (3.25) to (3.30) in Chap. 3.

These computed length changes can then be used to determine the velocity at the higher pressure by

$$\frac{v_P}{v_0} = \frac{f_P}{f_0}\frac{l_P}{l_0} \tag{7.9}$$

The actual changes in length and density computed in this way from data in Table 7.3 are shown in Table 7.4, and the velocities for the various modes can be computed at some selected higher pressure. When all the velocities are known at low and high pressures, the elastic constants can be computed

TABLE 7.4 Table of Density and Length Ratios

Condition	Ratio listed	Density ρ/ρ_0	Length l/l_0
$T = 23°C$	$\dfrac{\text{Value at } P = 30{,}000 \text{ lb/in}^2}{\text{Value at } P = 0}$	1.001305	0.99962
$T = -195.8°C$	$\dfrac{\text{Value at } P = 30{,}000 \text{ lb/in}^2}{\text{Value at } P = 0}$	1.001028	0.99966
$P = 0$	$\dfrac{\text{Value at } T = -195.8}{\text{Value at } T = 23°C}$	1.004525	0.99850

At $T = 23°C$ and $P = 0$ $\rho = 3.5833$ g/cm³
At $T = -195.8°C$ and $P = 0$ $\rho = 3.5995$ g/cm³
At $T = -195.8°C$ and $P = 30{,}000$ lb/in² $\rho = 3.6032$ g/cm³
At $T = 23°C$ and $P = 30{,}000$ lb/in² $\rho = 3.5880$ g/cm³

by linear combinations of the appropriate mode velocities: this is illustrated in col. 1 of Table 7.5. Of particular importance is the fact that this method allows a cross-check on each elastic constant. Agreement of 0.05 percent, as shown here, indicates that no errors have been made in orientation of

TABLE 7.5   Elastic Moduli for MgO (Units in kbar)

	$T = 23°C$			$T = -195.8°C$		
	$P = 1$ bar	$P = 2.0681$ kbar	dc/dP	$P = 1$ bar	$P = 2.0681$ kbar	dc/dP
$c_{11} = \rho v_1^2$	2,964.6	2,984.3	....	3,061.65	3,082.39	
$c_{11} = \rho(v_5^2 + v_3^2 - v_4^2)$	2,964.8	2,984.0	....	3,061.69	3,079.95	
Recommended	2,964.7	2,984.3	9.477	3,061.67	3,082.3	9.975
$c_{12} = \rho(v_1^2 - 2v_5^2)$	950.68	954.9	....	937.80	943.51	
$c_{12} = \rho(v_3^2 - v_4^2 - v_5^2)$	950.69	954.7	....	937.84	941.01	
Recommended	950.68	954.8	1.992	937.82	942.3	2.166
$c_{44} = \rho v_2^2$	1,558.7	1,561.3	....	1,576.04	1,578.8	
$c_{44} = \rho v_4^2$	1,559.1	1,561.3	....	1,575.62	1,578.8	
Recommended	1,558.9	1,561.3	1.160	1,575.84	1,578.8	1.431
Adiabatic bulk modulus $B_S = \frac{1}{3}(c_{11} + 2c_{12})$	1,622.0	1,631.3	4.497	1,645.8	1,655.6	4.738

the crystal or measurement of frequency. A further cross-check is seen in Table 7.2, where $v_7 = v_4$, and $v_8 = v_5$ within 0.1 percent. When such cross-checks are satisfied, one can be assured that the resulting value of the pressure derivative of the bulk modulus is good to within 1 percent. Thus, we see from Table 7.4 that the determination of B'_S for MgO gives the value 4.497 at 23° using the "chord method." This whole procedure for calculating *effective* pressure derivatives can be summarized by the example for c_{11}:

$$\frac{dc_{11}}{dP} = \frac{\rho v_1^2(\text{at } P) - \rho_0 v_1^2(\text{at } P = 0)}{\Delta P} \tag{7.10}$$

The second method, proposed by Thurston,[1] avoids computing the length change at higher pressure. Using the measured frequency derivative f', simple equations are derived in which the formula for the pressure derivative of the bulk modulus has no determination of length or density at the pressure. Thurston's formula for cubic crystals is

$$B'_S = 2c_{11}\frac{f'_1}{f_{10}} - \frac{4}{3}(c_{11} - c_{12})\frac{f'_5}{f_{50}} + \frac{1}{3}(1 + a\gamma T) \tag{7.11}$$

The last term in Eq. (7.11) is a correction which arises because the hydrostatic pressure is applied isothermally, while the ultrasonic waves are propagated adiabatically.

By substituting the data in Tables 7.3 and 7.4 into Eq. (7.11), it is found that $B'_S = 4.49$ at $23°C$. This method involves the tangents of the frequencies at zero pressure. Thus, in this example, the "tangent" method agrees with the "chord" method to the third significant figure. This confirms Eq. (7.11) and the assumption behind it that the relation between f and P is very close to a straight line.

Summary. If the second- and third-order coefficients are listed in the literature, B'_S is found by Eq. (7.5). If the pressure derivatives of the effective elastic constants are given, Eq. (7.7), which will have involved various computations such as Eq. (7.8), and a length correction, as given by Eqs. (3.25) to (3.30), are used. If the pressure derivatives of the frequency (the raw data) are available, Eq. (7.11) is used, where the notation for f_i is as given by Table 7.2.

7.3 Pressure Dependence of Elastic Constants of Hexagonal Crystals

Consider the hexagonal symmetry: *Laue group HI.* There are five independent second-order coefficients ($c_{11}, c_{12}, c_{13}, c_{33}, c_{44}$) and ten third-order coefficients ($C_{111}, C_{112}, C_{123}, C_{133}, C_{144}, C_{155}, C_{212}, C_{222}, C_{333}, C_{344}$).[7,8]

The pressure derivative of the bulk modulus can be given in terms of the pressure derivatives of effective moduli by

$$B'_S = \frac{1}{X}[(c_{11}+c_{12})c'_{33} + c_{33}(c'_{11}+c'_{12}) - 4c_{13}c'_{13} - B_S X'] \quad (7.12)$$

where
$$B_S = \frac{Z}{X} \quad (7.13)$$

$$X = c_{11} + c_{12} + 2c_{33} - 4c_{13} \quad (7.14)$$

$$Z = (c_{11}+c_{12})c_{33} - 2c_{13}^2 \quad (7.15)$$

Note that neither c_{44} nor its pressure derivative appears in the above equations. These equations also hold for trigonal symmetry since c_{14} does not affect the bulk modulus of crystals in this rhombohedral system.

In order to convert Eq. (7.12) to an expression involving the third-order constants, the relations between the pressure derivatives of effective and thermodynamic coefficients are needed, as well as the relation of third-order coefficients to the pressure derivatives of thermodynamic coefficients.

First, we obtain the relations between the effective and thermodynamic pressure derivatives.[3] The pertinent transformations are obtained from Table 7.1, noting that for hexagonal symmetry $s_1 = s_2$. They are

$$\frac{dc_{11}}{dP} + \frac{dc_{12}}{dP} = \frac{d\bar{c}_{11}}{dP} + \frac{d\bar{c}_{12}}{dP} + (s_3 - 2s_1)(c_{11} + c_{12}) \tag{7.16}$$

$$\frac{dc_{13}}{dP} = \frac{d\bar{c}_{13}}{dP} + 1 - s_3 c_{13} \tag{7.17}$$

$$\frac{dc_{33}}{dP} = \frac{d\bar{c}_{33}}{dP} - 1 + (2s_1 - 3s_3)c_{33} \tag{7.18}$$

where

$$s_1 = s_{11} + s_{12} + s_{13} \tag{7.19}$$

and

$$s_3 = s_{31} + s_{32} + s_{33} \tag{7.20}$$

Next we need the connection between thermodynamic pressure derivatives and the third-order coefficients. They have been worked out by Thurston.[4]

$$\bar{c}'_{11} + \bar{c}'_{12} = -[s_1(2C_{111} - C_{222} + 3C_{112}) + s_3(C_{113} + C_{123})] \tag{7.21}$$

$$\bar{c}'_{13} = -[s_1(C_{113} + C_{123}) + s_3 C_{133}] \tag{7.22}$$

$$\bar{c}'_{33} = -[2s_1 C_{133} + s_3 C_{333}] \tag{7.23}$$

Equations (7.16) to (7.23) are substituted in Eq. (7.12) to obtain B'_S as a function of third- and second-order elastic constants alone. From Eqs. (7.21) to (7.23), we see that only 7 of the 10 third-order elastic constants of a hexagonal crystal enter into the pressure derivative of the bulk modulus.

A formula must be developed that is analogous to Eq. (7.11) for hexagonal symmetry in order to show how B'_S can be determined from the primary frequency data. The modes are defined in terms of the directions in the crystal and the direction of the particle motion as shown in Table 7.6. Here the z axis is parallel to the c axis of the crystal and the x axis is parallel to one of the Bravias-Miller axes, a_1, a_2, a_3. The first step is to find the relationship between the effective-pressure derivatives and the frequency-pressure derivatives.[6] First, we will consider a cubic case, then generalize to hexagonal. The propagation velocity for the yth mode is

$$v_y = 2l f_y$$

and the appropriate elastic constant for that mode is

$$c_y = \rho v_y^2 \tag{7.24}$$

Thus

$$\rho v_y^2 = 4\rho l^2 f_y^2 \tag{7.25}$$

TABLE 7.6 Definitions for a Hexagonal Crystal

Designation	Mode	Direction* of propagation	Direction of particle motion	Modulus
v_1	Long	x axis	x	$c_{11} = \rho_0 v_1^2$
v_2	Shear	z axis	x or y	$c_{44} = \rho_0 v_2^2$
v_3	Long	z axis	z	$c_{33} = \rho_0 v_3^2$
v_4	Long	x axis	y	$c_{66} = \tfrac{1}{2}(c_{11} - c_{12}) = \rho_0 v_4^2$
v_5	Long	y axis	y	$c_{11} = \rho_0 v_5^2$
v_6	Shear	x axis	z	$c_{44} = \rho_0 v_6^2$
v_7	Shear	y axis	x	$c_{66} = \rho_0 v_7^2$
v_8	Quasi-long	$y'(45°$ to $z)$	$\perp x$	$c_{13} + c_{44} = 2[(\lambda_{22} - \rho v^2)(\lambda_{33} - \rho v^2)]^{1/2}$
v_9	Quasi-shear	$y'(45°$ to $z)$	$\perp x$	$\lambda_{22} = \tfrac{1}{2}(c_{11} + c_{44})$ $\lambda_{33} = \tfrac{1}{2}(c_{33} + c_{44})$

* The values x and z correspond to the a and c crystallographic axis; y is in the plane of the a axis and rotated 90° from a_1, so that a_1, y, and z form a cartesian coordinate set of axes.

Differentiating and combining the above,

$$\frac{dc_y}{dP} = c'_y = c_y \frac{2df_y}{f_y \, dP} + \frac{d\rho}{\rho_0 \, dP} + \frac{2dl}{l \, dP} \qquad (7.26)$$

Noting that $dl/l = -\tfrac{1}{3}(d\rho/\rho)$ for cubic symmetry, the above reduces to a simple form for cubic and isotropic solids: the initial slope at $P = 0$ becomes simply

$$(c'_y)_0 = (c_y)_0 \left(\frac{2f'_y}{f_{y0}}\right) + \frac{\chi_T}{3} \qquad (7.27)$$

Here χ_T is the isothermal compressibility equal to $1/\rho(d\rho/dP)_T$. For hexagonal symmetry the general formula replacing Eq. (7.27), but coming from Eq. (7.26), is

$$(c'_y)_0 = (c_y)_0 \left[\left(2\frac{f'_y}{f_{y0}}\right) + \chi_T - 2N_k N_m s_{kmi}\right] \qquad (7.28)$$

The last term in the above is the isothermal linear compressibility appropriate to the propagation direction y and is identified by the direction numbers N. We now proceed to identify the linear compressibility appropriate to each elastic constant.

Under hydrostatic pressure the strains along the z axis are not equal to the strains along the x axis in a hexagonal crystal. The equations for

180   Elastic Constants and Their Measurement

propagation are not invariant with pressure, as they are for the cubic case, but must reflect the anisotropic strains and their connection to path length.

Suppose a length l_0 in the xy plane at $P = 0$ becomes l/x at P, and a length l_0 along the z axis direction at $P = 0$ becomes l/z at P, where the compression factors x and z are equal to or greater than 1. With a given pressure, the path lengths and density change according to

$$l = l_0 x \tag{7.29}$$

$$l = l_0 z \tag{7.30}$$

$$l = \rho_0 x^2 z \tag{7.31}$$

The factors x and z are specified in terms of the isothermal constants from the theory of elasticity:

$$\frac{dx}{x} = \frac{(c_{33} - c_{13})\, dP}{c_{33}(c_{11} + c_{12}) - 2c_{13}^2} \tag{7.32}$$

$$\frac{dz}{z} = \frac{(c_{11} + c_{12} - 2c_{13})\, dP}{c_{33}(c_{11} + c_{12} - 2c_{13}^2)} \tag{7.33}$$

Examination of the propagation formulas in Table 7.6 with Eqs. (7.29), (7.30), and (7.31) leads to the following propagation formulas at pressure P^6:

$$\begin{aligned}
c_{11} &= z h_0 f_1^2 \\
c_{33} &= \frac{x^2}{z} h_0 f_3^2 \\
c_{44} &= \left(\frac{x^2}{z}\right) h_0 f_2^2 \\
c_{66} &= z h_0 f_4^2 \\
c_{11} &= z h_0 f_5^2 \\
c_{12} &= z h_0 (f_1^2 - 2 f_4^2) \\
c_{13} &= z h_0 \left\{ -f_6^2 + \left(\frac{x}{z}\right) [(2f_8^2 - f_3^2 - f_6^2)(2f_8^2 - f_2^2 - f_1^2)]^{1/2} \right\}
\end{aligned} \tag{7.34}$$

where $h_0 = 4\rho_0 l_0^2$.

Operating on Eq. (7.34) shows how Eq. (7.26) is to be expressed for each effective derivative. For example,

$$\left(\frac{dc_{11}}{dP}\right)_{P=0} = c'_{11} = (c_{11})\left(2\frac{f'_1}{f_{10}} + \frac{dz}{Z_0}\right)$$

$$c'_{33} = (c_{33})\left(2\frac{f'_3}{f_{30}} + 2\frac{dx}{x_0} - \frac{dz}{z_0}\right)$$

where x_0 and z_0 are unity. The strains dx and dz are isothermal and must be corrected to the adiabatic equivalents. We do this by noting that Eqs. (7.32) and (7.33) are expressions for the linear compressibility. Thus, they can be written as

$$\chi_T(\text{perp}) = \frac{c_{33} - c_{13}}{c_{33}(c_{11} + c_{12}) - 2c_{13}^2} \tag{7.35}$$

$$\chi_T(\text{para}) = \frac{c_{11} + c_{12} - c_{13}}{c_{33}(c_{11} + c_{12}) - 2c_{13}^2} \tag{7.36}$$

where $\chi_T(\text{perp})$ and $\chi_T(\text{para})$ are read as the linear isothermal compressibility at zero pressure perpendicular and parallel to the z axis. We note from Eq. (7.34) that $c_{ij}/h_0 z$ is a function only of the measurable frequencies at high pressure. Thus, if the numerators of Eqs. (7.32) and (7.33) are divided by $h_0 z$ and the denominators are divided by $(h_0 z)^2$, Eqs. (7.32) and (7.34) are replaced by

$$dx = \chi_T(\text{perp}) \cdot dP \tag{7.37}$$

$$dz = \chi_T(\text{para}) \cdot dP \tag{7.38}$$

Recalling that $\chi_T \chi_S = B_S/B_T = (1 + \alpha \gamma T)$ for a cubic case, the above can be transformed into an adiabatic formula by the factors

$$[1 + \gamma T \alpha(\text{perp})] = \frac{\chi_T(\text{perp})}{\chi_S(\text{perp})} \tag{7.39}$$

$$[1 + \gamma T \alpha(\text{para})] = \frac{\chi_T(\text{para})}{\chi_S(\text{para})} \tag{7.40}$$

where $\alpha(\text{perp})$ and $\alpha(\text{para})$ are the approximate linear coefficients of thermal expansion. Thus, we have

$$dx = [1 + \gamma T \alpha(\text{perp})]\chi_S(\text{perp}) \cdot dP \tag{7.41}$$

$$dz = [1 + \gamma T \alpha(\text{para})]\chi_S(\text{para}) \cdot dP \tag{7.42}$$

and by multiplying Eqs. (7.37) and (7.33) by the factors given by Eqs. (7.39) and (7.40), the various c_{ij} in Eqs. (7.35) and (7.33) can be replaced by measured frequencies prescribed by Eqs. (7.34). We now integrate Eqs. (7.41) and (7.42) to get x and z by expanding the compressibilities as a function of pressure with a McLaurin's series. We resort to the approximation of two terms in the expansion because the change of frequencies with pressure is small. For example,

$$\chi_S(\text{perp}) \simeq [\chi_S(\text{perp})]_0 - \chi_S'(\text{perp}) \cdot P \tag{7.43}$$

and similarly for the other compressibility. Equation (7.43) is evaluated numerically by placing the appropriate frequencies into Eqs. (7.32) and (7.33) and evaluating them at a higher pressure and a lower pressure. Finally, we have

$$x = 1 + [1 + \gamma T\alpha(\text{perp})][\chi_S(\text{perp}) \cdot P - \tfrac{1}{2}\chi_S'(\text{perp}) \cdot P^2] \quad (7.44)$$

$$z = 1 + [1 + \gamma T\alpha(\text{para})][\chi_S(\text{para}) \cdot P - \tfrac{1}{2}\chi_S'(\text{para}) \cdot P^2] \quad (7.45)$$

In most cases, especially for incompressible solids, the term in P^2 is so small that it can be ignored.

The next step is to form the appropriate combinations of pressure derivatives which enter into Eq. (7.12). The appropriate quantities are c_{33}, $c_{11} + c_{12}$, $c_{11}' + c_{12}' + 2c_{33}'$, and c_{13}'.

We have

$$c_{33}' = (c_{33})_0 \left[\left(2\frac{f_3'}{f_{30}} \right) + 2\left(\frac{dx}{dP} - \frac{dz}{dP} \right) \right] \quad (7.46)$$

$$c_{11}' + c_{12}' = 2(c_{11})_0 \left(2\frac{f_1'}{f_{10}} + \frac{dz}{dP} \right) - 2(c_{11})_0 \left(\frac{f_4'}{f_{40}} + \frac{dz}{dP} \right) \quad (7.47)$$

The expression for c_{13}' is found by differentiation of the last term in Eq. (7.34).

Thurston has worked out another expression for c_{13} at a higher pressure P. It is

$$c_{13} + c_{44} = -P + \sqrt{(Q_{22} - \rho v_8{}^2)(Q_{33} - \rho v_8{}^2)} \quad (7.48)$$

where $Q_{22} = \dfrac{c_{11} + c_{44}}{2}$

$Q_{33} = \dfrac{c_{33} + c_{44}}{2}$

Summary. If the effective pressure derivatives of the elastic constants are available, use Eq. (7.12). If the frequencies of the modes at high pressure (the basic data) are available, one can find the effective pressure derivatives by the "tangent" method by using Eqs. (7.45) and (7.46) along with (7.39) to (7.42), or one can evaluate the effective elastic constants at higher pressure by using Eqs. (7.34) and (7.47) along with (7.43) and (7.44) and then form a difference to be divided by the pressure. This is the "chord" method. If the second- and third-order coefficients are available, Eqs. (7.16) to (7.23) are used.

The equations relating the pressure derivatives of the modes to the third-order and second-order constants have been presented by Brugger.[9]

We terminate this section by mentioning that one case of trigonal symmetry has been completely worked out, namely, the case of α-quartz.[10–12]

If the pressure derivatives of the effective elastic constants are given, B'_S is found by using Eq. (7.12). However, the expression for the third-order constants is more complicated for the hexagonal case, since there are 14 third-order coefficients for α-quartz.

7.4 Assorted Formulas for Young's Modulus, Shear Modulus, and Poisson's Ratio

The pressure derivative of the bulk modulus has been emphasized in the previous section. However, if the modes are completely determined as they vary with pressure, as shown in Table 7.3 for MgO, the pressure variation of elastic moduli other than the bulk modulus can be defined. In this section, we will systematize formulas for Young's modulus, shear modulus, and Poisson's ratio for crystals of cubic symmetry.

The definition corresponding to Eq. (7.4) is

$$E_S = c_{11} - \frac{2(c_{12})^2}{c_{11} + c_{12}} \tag{7.49}$$

where E_S represents Young's modulus in the [100] direction at ambient pressure. We note that E depends upon the direction in the crystal, but here we shall deal only with the [100] direction. Defining the elastic constant mode $w_i = \rho v_i^2$, where v_i is given by the definitions in Table 7.2, we have in place of Eq. (7.49) the definition of Young's modulus:

$$E_S = \frac{w_5(3w_1 - 4w_5)}{w_1 - w_5} \tag{7.50}$$

$$E_S = \frac{w_5(3w_3 - w_5 - 3w_4)}{w_3 - w_4} \tag{7.51}$$

in terms of modes 3, 4, and 5.

For a uniaxial stress along an arbitrary direction specified with respect to the crystal axes by the direction cosines l_1, m_1, n_1, Young's modulus for either isothermal or adiabatic conditions satisfies

$$\frac{1}{E_S} = s_{11} + (s_{44} + 2s_{15} - 2s_{11})(l_1^2 m_1^2 + m_1^2 n_1^2 + n_1^2 l_1^2) \tag{7.52}$$

where quantities s_{ij} are the conventional compliance coefficients. The lateral contraction resulting from the uniaxial stress depends on the direction of stress and direction under consideration. In the case of isotropy, where $s_{44} = 2(s_{11} - s_{12})$, Eq. (7.52) reduces simply to

$$\frac{1}{E_S} = s_{11} \tag{7.53}$$

The pressure derivative of Young's modulus (in the [100] direction) is thus, using Eqs. (7.50) and (7.51),

$$E'_S = \frac{[w'_5(3w_1{}^2 + 4w_5{}^2 - 8w_5 w_1) + w_5{}^2 w'_1]}{(w_1 - w_5)^2} \tag{7.54}$$

or $\quad E'_S = \dfrac{w'_5(3w_3{}^2 - 6w_3 w_4 - 2w_3 w_5 + 2w_4 w_5 + 3w_4{}^2) + w_5{}^2(w'_3 - w'_4)}{(w_3 - w_4)^2} \tag{7.55}$

Equations (7.54) and (7.55) involve corrections for length under pressure in order to determine w'_5 and w'_1. This requires applying the corrections outlined in the preceding section. The formulas corresponding to the "tangent" method have been worked out to the point where one uses only the frequency-pressure derivatives as given in Table 7.3. In this case Eq. (7.54) is given by

$$E'_S[(w_1 - w_5)^2] = w_5(3w_1{}^2 + 4w_5{}^2 - 8w_5 w_1)\frac{2f'_5}{f_{50}} + \frac{1 + \alpha\gamma T}{3B_S}$$

$$+ w_5{}^2 w_1 \left(\frac{2f'_1}{f_1}\right) + \frac{1 + \alpha\gamma T}{3B_S} \tag{7.56}$$

We now turn to Poisson's ratio for a cubic crystal and consider the [100] direction.

$$\sigma_S = \frac{c_{12}}{c_{11} + c_{12}}$$

In terms of the w_i modes, the above is either

$$\sigma_S = \frac{w_1 - 2w_5}{2(w_1 - w_5)} \tag{7.57}$$

or $\quad \sigma_S = \dfrac{w_3 - w_5 - w_4}{2(w_3 - w_4)} \tag{7.58}$

A more general formula for Poisson's ratio has been given by Thurston in which the subscripts on σ refer to the cosine directions (ratio of lateral strain along direction 2 to extensional strain along direction 1):

$$\sigma_{21} = -E\left[s_{12} + \left(s_{11} - s_{12} - \frac{s_{44}}{2}\right)(l_1{}^2 l_2{}^2 + m_1{}^2 m_2{}^2 + n_1{}^2 n_2{}^2)\right] \tag{7.59}$$

In this formula, E is the direction-dependent E given by Eq. (7.52). These formulas hold with either adiabatic coefficients on both sides or with isothermal coefficients on both sides.

Returning to σ for the [100] direction, the appropriate pressure derivative is given by

$$\sigma'_S = \frac{w_5 w'_1 - w_1 w'_5}{2(w_1 - w_5)^2} \tag{7.60}$$

or

$$\sigma'_S = \frac{w_5(w'_3 - w'_4) + (w_4 - w_3)w'_5}{2(w_3 - w w_4)^2} \tag{7.61}$$

The pressure derivative for the above given in terms of the frequency-pressure derivative is

$$\sigma'_S = \frac{w_1 w_5 [(f'_1/f_{10}) - (f'_5/f_{50})]}{(w_1 - w_5)^2} \tag{7.62}$$

or

$$\sigma'_S = \frac{w_4 w_5 [(f'_5/f_{50}) - (f'_4/f_{40})] - w_3 w_5 [(f'_5/f_{50}) - (f'_3/f_{30})]}{(w_3 - w_4)^2}$$

The shear modulus in the [100] direction is given simply as

$$G = c_{44} = w_4 \tag{7.63}$$

The pressure derivative of the above is

$$G' = w'_4 \tag{7.64}$$

$$G' = w_4 \left[2\frac{f'_4}{f_4} + \left(1 + \frac{\alpha \gamma T}{3 B_S}\right) \right] \tag{7.65}$$

No subscript is given for G, since the isothermal value is equal to the adiabatic value. As long as the pressure derivative of the shear modulus is taken isothermally, as it is by our prime, the shear modulus can be considered either in the isothermal or adiabatic form. However, care must be exerted if one wishes to compare the pressure derivative taken adiabatically with that taken isothermally.

It is of some interest to compute the value of G' for an isotropic material (which is what one would expect to measure on a polycrystalline material) given the pressure derivatives of the elastic constants of a single crystal. To do this, we use the Voigt-Reuss method (see Chap. 2). It is extensively discussed in the literature, but only two recent references will be given.[13,14] One should take a value between the two following limits for cubic crystal data.

Voigt:
$$G' = \tfrac{1}{5}(c'_{11} - c'_{12}) + \tfrac{3}{5} c'_{44} \tag{7.66}$$

Reuss:
$$G' = \frac{4}{5}\left(\frac{a}{c_{11} - c_{12}}\right)^2 c'_{11} - c'_{12} + \frac{3}{5}\left(\frac{a}{c_{44}}\right)^2 c'_{44} \tag{7.67}$$

where
$$a = \frac{5 c_{11} c_{44} - 5 c_{12} c_{44}}{3 c_{11} - 3 c_{12} + 4 c_{11}} \tag{7.68}$$

For the hexagonal case, a number of combinations of elastic constants arise. In addition to X and Z, defined by Eqs. (7.14) and (7.15), the following combinations are useful:

$$B_V = \tfrac{1}{3}(2c_{11} + 2c_{12} + c_{33} + 4c_{13}) \tag{7.69}$$

$$X = Z(c_{44} + c_{66}) + 3B_V c_{44} c_{66} \tag{7.70}$$

Equation (7.69) has physical significance. It is the bulk modulus defined according to the Voigt approximation. This definition is not the same as the one customarily used in thermodynamic descriptions of hexagonal solids given by Eq. (7.13). The latter equation corresponds to the Reuss definition in the Reuss-Voigt schemes.

The limiting values of the pressure derivatives of isotropic media calculated from the elastic-constant data obtained on a single crystal are

Voigt: $\quad G' = \tfrac{1}{30}(E' + 12c'_{44} + 12c'_{66})$ \hfill (7.71)

Reuss: $\quad G' = 6 \left[\dfrac{5}{4} B_V \left(\dfrac{c_{44} c_{66}}{X} \right)^2 Z' - \dfrac{5}{4} \left(\dfrac{c_{44} c_{66}}{X} \right)^2 B'_V \right.$

$\left. + \dfrac{5}{12} \left(\dfrac{Z c_{11} c_{12}}{c_{11} - c_{12}} \right)^2 (c'_{11} - c'_{12}) + \dfrac{5}{12} \left(\dfrac{Z c_{11} c_{66}}{c_{44}} \right)^2 c'_{44} + \dfrac{5}{12} (Z c_{11})^2 c'_{66} \right]$

(7.72)

where Z is defined by Eq. (7.15) and X by Eq. (7.14).

It is significant that c'_{44} and c'_{66} can be negative, as in the case of CdS recently reported by Corll.[15] He found $c'_{11} = 3.059$, $c'_{33} = 3.22$, $c'_{13} = 4.616$, $c'_{12} = 4.75$, $c'_{44} = 0.835$, and $c'_{66} = -0.627$. Thus the limit of G', according to Eqs. (7.71) and (7.72), is negative, becoming about -0.72, a value that one may expect to be reasonably close to polycrystalline CdS. Polycrystalline ZnO was measured by Soga and Anderson,[16] and the measured value of G' was reported as -0.69.

If one generalizes to the trigonal crystals, the expressions of the Voigt-Reuss bulk modulus remain the same, as mentioned. The expression for the trigonal Reuss value of G' corresponding to Eq. (7.72) is complicated and will not be reproduced here. The expression for the Voigt value of G' for trigonal crystal is the same as in Eq. (7.72). The pressure derivatives of the α-quartz moduli have been reported by McSkimin et al.[11] They are $c'_{11} = 3.28$, $c'_{12} = 8.66$, $c'_{13} = 5.97$, $c'_{14} = 1.93$, $c'_{33} = 10.84$, $c'_{44} = 2.86$, $c'_{66} = 2.69$. Computing G' from these values, one expects it to be close to zero, perhaps slightly negative, for the ideal polycrystalline solid. Soga[18] computed the value of the pressure derivative of the shear velocity and found it to be negative.

7.5 Adiabatic Isothermal Transformations

The equations presented in the previous sections show how the pressure derivatives of elastic constants can be found from ultrasonic experiments. These equations yield adiabatic values. When these values are to be compared with experimental results arising from measurements made under purely isothermal conditions, conversions have to be made. What is required, for example, are the relations between B'_T and B'_S; that is, the pressure derivatives are both isothermal but the moduli are different.

The fundamental equation is given by

$$B_S = B_T(1 + \alpha\gamma T) \tag{7.73}$$

Using the fact that $E_T = 1/s_{11}$, and $s_{11} = s_{11} + T\alpha^2/\rho C_P$ and the definition of γ given by Eq. (7.6), one obtains for Young's modulus

$$E_S = E_T\left[1 + \left(\frac{\alpha\gamma T}{B_S}\right)\right] \tag{7.74}$$

$$= E_T(1 + \Delta) \tag{7.75}$$

Similarly, for Poisson's ratio, since $\sigma_T = s_{12}/s_{11}$, we find that

$$\sigma_S(1 - \Delta) = \sigma_T(1 + \Delta) \tag{7.76}$$

Manipulation of the above equations leads to pressure derivatives. In the case of the bulk modulus, we have

$$B'_T = \frac{a^2}{b^2}\left[B'_S - \frac{2b^2}{a\alpha B_S}\left(\frac{\partial B_S}{\partial T}\right)_P + \frac{b^2}{2}\left(\frac{\partial \alpha}{\partial T}\right)_P\right] + b\left[1 + 2\gamma + 2T\left(\frac{d\gamma}{dT}\right)_P\right] \tag{7.77}$$

where $b = \alpha\gamma T$
$a = b/(1 + b)$

The right-hand side of Eq. (7.77) contains expressions which arise from adiabatic moduli measured versus pressure and temperature. In addition, there are terms requiring the variation of expansivity with temperature and the Grüneisen constant with temperature.

Expressions needed to evaluate Young's modulus and Poisson's ratio (in the [100] direction) are

$$\frac{d}{dP}(s_{11})_0 = \frac{E'_S}{(E_S)^2} + \frac{m^2}{b}\left[\frac{2(1+b)}{\alpha B_S}\left(\frac{\partial B_S}{\partial T}\right)_P - r\right] \tag{7.78}$$

$$\frac{d}{dP}(s_{12})_0 = \frac{E'_S}{(E_S)^2} - \frac{1}{E_S}\left(\frac{\partial \sigma_S}{\partial P}\right)_T + \frac{m^2}{b'}\left[\frac{2(1+b)}{\alpha B_S}\left(\frac{\partial B_S}{\partial T}\right)_P - r\right] \tag{7.79}$$

where
$$m = \frac{\alpha \gamma T}{3B_S} \tag{7.80}$$

and
$$r = \left(\frac{b}{a^2}\right)\left(\frac{\partial \alpha}{\partial T}\right)_P + 2T\left(\frac{\partial \gamma}{\partial T}\right)_P + 2b + 3 + 2\gamma \tag{7.81}$$

$$b = (\alpha \gamma T) \quad \text{and} \quad a = \frac{b}{1+b} \tag{7.82}$$

Finally,
$$E'_T = (E_T)^2 \frac{d(s_{11})_0}{dP} \tag{7.83}$$

and
$$\sigma'_T = \frac{[(\sigma_S/E_S) - (T\alpha^2/\rho C_P)]s'_{11} + [(1/E_S) + (T\alpha^2/\rho C_P)]s'_{12}}{(1/E_S + T\alpha^2/\rho C_P)} \tag{7.84}$$

We remind the reader, in passing, that $G'_S = G'_T$, where the pressure derivative is understood to be taken isothermally.

Of special interest are the cases for isotropic solids, where Young's moduli and Poisson's ratio are given in terms of derivatives of the bulk modulus and shear modulus.

$$E'_T = 9\left\{\frac{G^2 B'_T + [3B_S/(1+\alpha\gamma T)]^2 (\partial G/\partial P)_T}{[3B_S/(1+\alpha\gamma T) + G]^2}\right\} \tag{7.85}$$

$$\sigma'_T = 18\left\{\frac{G^2 B'_S - [B_S/(1+\alpha\gamma T)]^2 (\partial G/\partial P)_T}{[6B_S/(1+\alpha\gamma T) + 2G]^2}\right\} \tag{7.86}$$

We now turn to the transformations between adiabatic and isothermal temperature derivatives, that is, for example, relations between $(\partial B_T/\partial T)_P$ and $(\partial B_S/\partial T)_P$. The main equation can be found by substituting the volume expansivity α for compressibility in Eq. (7.27). Thus

$$(c'_y)_0 = (c_y)_0 \left(\frac{2f'_y}{f_{y0}} - \frac{\alpha}{3}\right) \tag{7.87}$$

where the primes now refer to derivatives with regard to temperature. We have

$$\left(\frac{\delta B_T}{\partial T}\right)_P = \frac{a}{b}\left\{\left(\frac{\partial B_S}{\partial T}\right)_T - \alpha B_S \frac{a}{b}\left[\gamma + \frac{b}{a^2}\left(\frac{\partial \alpha}{\partial T}\right)_P + T\left(\frac{\partial \gamma}{\partial T}\right)_P\right]\right\} \tag{7.88}$$

$$\left(\frac{\partial E_T}{\partial T}\right)_P = 9\left\{\frac{9(B_S)^2(\partial E_S/\partial T)_P + bE_S(\partial B_S/\partial T)_P - q}{(9B_S + bE_S)^2}\right\} \tag{7.89}$$

$$\frac{\partial \sigma_T}{\partial T} = 9\left(\frac{9(B_S)^2(\partial \sigma_S/\partial_T)_P - b(1+\sigma_S)\{h + E_S[q + (\partial \sigma_S/\partial T)_P]\}}{(9B_S + bE_S)^2}\right) \tag{7.90}$$

where $q = \alpha B_S \left[\dfrac{1}{\alpha T} + \left(\dfrac{1}{\alpha \gamma}\right)\left(\dfrac{\partial \gamma}{\partial T}\right)_P + \left(\dfrac{1}{\alpha^2}\right)\left(\dfrac{\partial \alpha}{\partial T}\right)_P \right]$

$h = B_S \left(\dfrac{\partial E_S}{\partial T}\right)_P - E_S \left(\dfrac{\partial B_S}{\partial T}\right)_P$

These equations provide the means of calculating the values of the isothermal pressure and temperature derivatives from the values determined in laboratory measurements, using ultrasonic measuring techniques which yield the adiabatic values of the pressure and temperature derivatives. The isothermal values are the ones needed in many applications, for example, the parameters B_0 and B'_0 needed in equation of state calculations. Thus, the values determined from ultrasonic techniques must first be transformed using the equations presented above.

REFERENCES

1. Thurston, R. N.: Ultrasonic Data and the Thermodynamics of Solids, *Proc. IEEE*, **53**:1320 (1965).
2. Thurston, R. N., and K. Brugger: Third-Order Elastic Constants and Velocity of Small Amplitude Elastic Waves in Homogeneously Stressed Media, *Phys. Rev.*, **133**: A1604 (1964).
3. Thurston, R. N.: Effective Elastic Coefficients for Wave Propagation in Crystals under Stress, *J. Acoust. Soc. Am.*, **37**:348 (1965).
4. Thurston, R. N.: Calculation of Lattice Changes with Hydrostatic Pressure from Third-Order Elastic Constants, pt. 2, *J. Acoust. Soc. Am.*, **41**:1093 (1967).
5. McSkimin, H. J.: Elastic Moduli of Single Crystal Germanium as a Function of Hydrostatic Pressure, *J. Acoust. Soc. Am.*, **30**:314 (1958).
6. Cook, R. J.: Variations of Elastic Constants and Static Strains with Hydrostatic Pressure: A Method for Calculation from Ultrasonic Measurements, *J. Acoust. Soc. Am.*, **29**:445 (1957).
7. Brugger, K.: Thermodynamic Definition of Higher Order Elastic Coefficients, *Phys. Rev.*, **133**:A1611 (1964).
8. Brugger, K.: Third-Order Elastic Coefficients in Crystals, *J. Appl. Phys.*, **36**[3]:759 (1965).
9. Brugger, K.: Determination of the Third-Order Elastic Constants in Crystals, *J. Appl. Phys.*, **36**:767 (1965).
10. McSkimin, H. J.: Measurement of the 25°C Zero-Field Elastic Moduli of Quartz by High-Frequency Plane-Wave Propagation, *J. Acoust. Soc. Am.*, **34**:1271 (1962).
11. McSkimin, H. J., P. Andreatch, Jr., and R. N. Thurston: Elastic Moduli of Quartz Versus Hydrostatic Pressure at 25°C and − 195.8°C, *J. Appl. Phys.*, **36**:1624 (1965).
12. Thurston, R. N., H. J. McSkimin, and P. Andreatch, Jr.: Third-Order Elastic Coefficients of Quartz, *J. Acoust. Soc. Am.*, **37**:267 (1966).
13. Hill, R.: Elastic Behaviour of a Crystalline Aggregate, *Proc. Phys. Soc. Lond.*, **65A**:350 (1952).
14. Anderson, O. L.: A Simplified Method for Calculating the Debye Temperature, *J. Phys. Chem. Sol.*, **24**:909 (1963).

15. Corll, J. A.: Effect of Elastic Parameters and Structure of CdS, *Phys. Rev.*, **157**:623 (1967).
16. Soga, N., and O. L. Anderson: Anomalous Behavior of the Shear Sound Velocity Under Pressure for Polycrystalline ZnO, *J. Appl. Phys.*, **38**:2985 (1967).
17. Soga, N., Temperature and Pressure Derivatives of Isotropic α-Quartz, *J. Geophys. Res.*, **73**: 827(1968).

Index

Absorption band, 158
Acoustic impedance, 48
Acoustic interferometry, 59
Acoustic loss, 121
  (See also Q)
Acoustic spectrum, 136
Acoustics, 5, 9, 169
Adiabatic character, 31, 76, 181
Adiabatic to isothermal transformations, 32, 33, 87, 188
Anelastic properties, 83
Anelasticity, 120
Angular coordinates, 128
Anisotropic body, 102
Anisotropic materials, 3, 28, 33, 37, 40
Anisotropy, 140
Atomic weight, mean, 144
Axes, orthogonal, 10
Axial transformation, 18

Bessel's function, 129

Bond:
  lapped, 74
  quality of, 74
  transducer to specimen, 61
Bond corrections, 59
Born model, 149
Buffer rod, 64
Bulk modulus, 4–6
  calculation: from other moduli, 6, 30, 108, 169
    from velocities, 6, 76
  estimation from infrared reflectivity, 157
  estimation from volume, 149
    pure compounds, 149
    solid solutions, 156
  linearity with pressure, 174
  pressure derivatives, 76, 170, 176
    estimation of, 163–166
  (*See also* Modulus)

Cauchy relation, 8

Index

Center of symmetry, 8, 19
Chain rule, 155
Coefficient:
　piezocaloric, 32
　reflection, 48
　of transformation, 18, 104
　of transmission, 48
Compliance (see Elastic compliance)
Composite oscillator, 115
Compressibility, 33, 76, 157, 175
　isothermal, 179
　linear, 179
Constructive interference, 69
Coordinate system, right-handed, 17
Crystal:
　classes, 14, 16, 17
　form, 16
　measurements on, 103–107, 174–182
　polycrystalline, 71
Crystal symmetry, 7, 14–15
Crystal systems, 15–18
　cubic, 7, 14–17, 23, 28, 31, 37, 104, 170, 176, 183
　hexagonal, 15, 17, 24, 177–182, 186
　　minerals, 106
　monoclinic, 14, 15, 19, 20
　orthorhombic, 15, 17, 21, 22
　tetragonal, 14–17, 21, 22
　　minerals, 106
　triclinic, 14–17, 19, 28
　trigonal, 15, 17, 24, 182, 186
Curie point, 55
Cylindrical rods, 90

Debye approximation, 160
Debye temperature, 144, 147, 160
Deformation:
　elastic, 2, 11, 31
　plastic, 2
Delay, trigger-point, 52
Delay line, mercury, 38, 51
Derivatives of elastic moduli, 76, 168–186
Determinant, 28
Diatomic solids, 158
Dielectric constant, 158
Diffraction, 46
Dilatation, 5
Direction cosines, 18, 20, 27, 104
Discriminator circuit, 54
Dispersion, 46
Displacement, particle, 128
Distortion, 5
Dynamic resonance, 88–124
　excitation methods for, 99, 114–120
　measuring system for, 98

Eddy currents, 117
Elastic behavior, 13
Elastic body, 127
Elastic compliance, 13, 31, 104, 112
　adiabatic, 33
　isothermal, 33
Elastic constants, 2–32, 126
　cross-checks, 173, 176
　effective, 171
　estimation, 143
　isotropic, 7
　matrix, 24
　nonisotropic, 7
　pressure derivatives of, 171
　second-order, 177
　tensor, 19
　thermodynamic, 171
　third-order, 170, 177
Elastic properties, 1
Elastic stiffness, 13, 31, 32
Elasticity theory, 4
Ellipticity, 138
Energy, internal, 154
Entropy, 32
Excitation methods, 99, 114–120
Equation:
　of motion, 27
　of state, 166, 168

First arrival of acoustic energy, 37
Forces:
　central, 8
　dissipative, 31
　inertial, 5
Form, crystal, 16
Fourier transforms, 46
Frequency:
　carrier, 59
　nondimensional, 131, 139
　pulse-repetition, 59
　ratio, 63, 108, 139
　resonant, 83, 131

Gating circuit, 53, 63
Generator:
　harmonic, 63
　pulse, 36, 40, 52
　time mark, 51
Grüneisen parameter, 76, 163
　first and second, 159
　pressure derivative of bulk modulus from, 163

Index

Harmonic generator, 63
Heat capacity, 32, 76
  (*See also* specific heat)
Hexagonal symmetry, 15, 17, 24
  calculation of elastic constants, 177–182
Hooke's law, 4, 13, 26, 103

Inertia:
  lateral, 90
  moment of, 119
Inertial bar, 113
Infra-red reflection (*see* Bulk modulus)
Interference, constructive, 69
Internal friction, 107, 114, 117, 120
Isotropic body, 88, 102, 127
  calculation of Young's modulus for, 102
Isotropic materials, 3–8, 14, 23, 25, 26, 28, 33, 37, 71, 106

Lamé constant, 6, 25, 37, 127
Lissajous patterns, 99
Longitudinal mode, 170–171
Longitudinal vibration, 82, 83, 88, 91
Longitudinal wave, 5, 45
Loss, acoustic, 55
  (*See also* Internal friction)

Madelung constant, 149
Materials, nonisotropic, 3, 28, 33, 37, 40
  acoustic coupling, 47, 59, 73, 74
  polycrystalline, 19, 30, 71, 127
Matrix:
  compliance, 14, 171
  elastic constants, 14
  inversion of, 14
  notation, 13
  reduced for crystal classes, 18–25
  reduction of, 14, 17
  transformation, 18
Matter tensor, 12
Maxwell's equations, 154
Mean atomic weight, 144
Measurements (*see* Method of measurement)
Mechanical wave, 26
Medium:
  anisotropic (*see* Materials, nonisotropic)
  bounded, 45
  semi-infinite, 26

Method of measurement:
  differential path, 56–58
  dynamic resonance, 88–124
  electrodynamic effect, 118
  electromagnetic effect, 117
  electrostatic effect, 119
  forced-vibration, 120
  free-vibration, 120
  interferometric, 59–73
  long-pulse, 64
  phase-comparison, 59–68
  piezoelectric effect, 115
  pulse-echo, 46–58
  pulse-superposition, 68–73
  pulse-transmission, 36–44
  sing-around, 52
  sphere resonance, 126–142
  transit-time (*see* pulse transmission *above*)
  vibration of wire, 111
Methods, static, 83
  of excitation, 99
Mirror plane of symmetry, 16–22
Mode:
  conversion, 37, 43
  effect of ellipticity, 138
  flexural vibration, 88, 100
  fundamental vibration, 127
  identification, 100, 136
  longitudinal, 171
    vibration, 91
  poloidal, 127
  spheroidal, 127, 139
  toroidal, 127
  torsional vibration, 84, 100, 116, 126, 139
Modulus:
  adiabatic, 32, 33, 76, 169
  bulk, 4, 6, 25, 28, 30, 33, 37, 77, 147, 154, 161, 169
    Reuss approximation, 30
    Voigt approximation, 30
    (*See also* Bulk modulus)
  elastic, 2–7, 30, 77, 82, 116
    calculation from resonant frequencies, 101
    estimation of temperature derivatives, 159–162
    free, 103
    isotropic, 147
    pure, 103
  isothermal, 32, 33, 169
  isotropic, 28
  longitudinal, 5, 25, 68
  shear, 4–8, 25, 28, 30, 37, 77, 102, 113, 127, 161, 183
    effect of orientation on, 102

Modulus, shear (*Cont.*):
  free, 105
  pressure derivative of, 77, 183
  Reuss approximation, 30
  temperature dependence, 108
  Voigt approximation, 30
Young's, 2–7, 25, 37, 88, 90, 101–111, 147, 161, 167, 183
  effect of orientation on, 102
  free, 105
  pressure derivatives, 183
  temperature dependence, 108, 159, 188
Molar volume, 156
Motion:
  equations of, 27
  particle, 173
  simple harmonic, 60, 113

Newton's laws, 5
Nodal surface on sphere, 128
Node, 100
  positions of, 101

Orthogonality, 18
Oscillation:
  flexural, 88
  free, 127
  longitudinal, 91
  sphere, 127
  spheroidal, 130
  toroidal, 137
  vibrational, 84
Oscillator, composite, 115
Overtone, 84, 89

Particle displacement, 128
Particle motion, 173
Phase angle, 67
  calculation of, 67
Phase comparison 59, 62–67, 77
  double-pulse, 59
Phase-shift, 61
Piezocaloric coefficient, 32
Plane wave, 25
Point group symmetry operators, 16
Poisson's ratio, 2–8, 25, 44, 77, 88, 90, 101, 130, 136, 145, 158, 161
  pressure derivatives of, 185
Polycrystalline, 71
Pressure, 4, 75
Pressure dependence:
  cubic crystals, 170–176

Pressure dependence (*Cont.*):
  hexagonal crystals, 177–182
  polycrystals, 75–77
Pressure derivatives, 77, 169, 183
  chord method, 75–77, 176, 182
  effective, 176, 178, 182
  estimated, 163–166
  tangent method, 177, 182
  thermodynamic, 178
Propagation direction, 28
  in cubic crystals, 174
  in hexagonal crystals, 179
Property tensor, 12
Pulse:
  direct-current (dc), 46
  generator, 36, 40, 52
  oscillator, 50, 52, 57, 68
  radio-frequency (rf), 46
  shape, 46–50
  superposition, 59, 68–71
  transmission, 43, 58
Pulse-echo technique, 46, 50, 55, 58

Q, 58, 136
Quasi-isotropic, 29

Radial function, 128
Radius of gyration, 85, 89, 111
Rectangular bars, vibration of, 83
  flexural, 90
  longitudinal, 91
  torsional, 84
Reflection, acoustic wave, 46
Reflection coefficient, 48
Refractive index, 158
Repetition frequency, 50
  pulse-(PRF), 68
Repulsion potential, 149
Resonance:
  bar, 42, 83
  cylinder, 83
  dynamic, 82–124
  flexural, 86–90
  longitudinal, 91
  sphere, 126–142
  torsional, 84
Resonant condition, 112
Resonant frequency, 83, 127
  critical, 121
  flexural, 89, 107
  longitudinal, 91
  temperature dependence of, 108
  torsional, 84, 89, 107
Reststrahlen frequency, 158

Reuss approximation, 30, 185
Rotary inversion, axis, 16
Rupture, 2

Seal, wrung-on, 74
Seal corrections, 59
Secular equation, 28
Semi-infinite body, 26
Shape factor, 84–89, 102
Shear modulus (see Modulus, shear)
Shear strain, 3, 11
Shear stress, 3, 10
Shear velocity, 6, 28, 37, 42, 44, 131, 137, 139
  effect of orientation on, 102
  estimated from specific heat, 144–146
  pressure derivatives, 77, 185
  from vibrating bar, 84
  from vibrating sphere, 131, 136–139
Simple harmonic motion, 60, 113
Sing-around method, 52–54
Solid:
  covalent, 8
  diatomic, 158
  incompressible, 182
  ionic, 8, 149
  isotropic, 37
Solid solution, bulk modulus of, 156
Specific heat, 144, 147
  (See also Heat capacity)
Specimen, small, measurements on, 59, 71, 118, 126
Sphere, resonance, 126–142
  data analysis, 136
  equipment for, 133
  method of operation, 135
  mode identification, 136
  specimen preparation, 132
    grinder for, 132
Spheroid, 137
Square bars, measurement of, 91
Standing waves, 82
Stiffness (see Elastic stiffness)
Strain, 1–13, 25
  adiabatic, 32
  components of, 12, 26
  energy, 11
  isothermal, 32
  reversible, 2
Stress, 1–13, 25
  components of: normal 10, 26
    shear, 10
  compressional, 4
  inhomogeneity, 114
  level, 2
  shear, 4, 10

Stress (*Cont.*):
  tensile, 3, 14
  uniaxial, 3, 183
Szigeti equation, 158

Temperature, 4, 37
  characteristic, 160
  Debye, 144, 147, 160
  dependence of moduli, 159–162
Temperature derivatives, 159–162, 168, 188
Tensors:
  compliance, 13, 171
  fourth-rank, 13, 18
  matter, 12, 15
  notation, 18
  property (see matter *above*)
  second-rank, 10
  stiffness, 13
  strain, 12–14
  stress, 12–14, 26
  symmetric, 12
  transformation, 18, 104
    elements of, 18–20, 104
Thermal expansivity, 32, 64, 170, 181
Thermodynamics, 4, 32, 169
  variables, 155
Thin films, 158
Time of flight, 58
Time delay, 42, 52, 55, 62, 69
  change in, measurement of, 54–57, 62–64
Time mark generator, 51
Tone burst generator, 46
Transducer seal boundary, 61
Transducers, 38, 59, 73
  bond, 61
  high-temperature, 122, 140
  magnetostrictive, 118
  quartz, 38, 116
  seal, 61
Transformation:
  adiabatic to isothermal, 32
  of pressure derivatives, 187–188
  axes, 18, 104
  coefficients, 18, 104
  matrix, 18, 104
Translation, 11

Velocimeter, 38
Velocity:
  bar, 37
  longitudinal, 7, 37, 42, 130
  mean, 144
  Rayleigh wave, 45

Velocity (*Cont.*):
  shear (*see* Shear velocity)
  sound, 9, 61, 63, 70
Vibration:
  flexural, 82, 88, 100
  free, 83
  fundamental, 88, 90
  of infinite bar, 85, 88
  longitudinal, 82, 83, 88, 91
  modes of, 8, 100
  optical, 144
  rotary, 129
  torsional, 82, 84, 100, 116, 137
Voigt approximation, 30, 185
Voigt-Reuss-Hill approximation, 30

Wave:
  dilatational, 37
  longitudinal, 37, 38, 42, 44

Wave (*Cont.*):
  normal, 27
  Rayleigh, 45
  shear, 42, 44
  standing, 82
  velocity of, 82
Wave equation, 5
Wave propagation, 25

Young's modulus, 2–7, 25, 37, 88, 90, 101–111, 147, 161, 167, 183
  from cylindrical rods, 90, 91
  effect of orientation on, 102
  example calculation, 101
  free, 105
  pressure derivatives, 184
  from rectangular bars, 90
  from square bars, 91
  temperature dependence, 108, 159–162, 188